Modern Particle Physics: Concepts and Applications

Modern Particle Physics: Concepts and Applications

Aidan Butler

STATES
ACADEMIC PRESS
www.statesacademicpress.com

Published by States Academic Press,
109 South 5th Street,
Brooklyn, NY 11249, USA

ISBN: 978-1-63989-363-8

Cataloging-in-Publication Data

Modern particle physics : concepts and applications / Aidan Butler.
p. cm.
Includes bibliographical references and index.
ISBN 978-1-63989-363-8
1. Particles (Nuclear physics). 2. Nuclear physics. 3. Physics. I. Butler, Aidan.
QC793.2 .C66 2022
539.7--dc23

For information on all States Academic Press publications
visit our website at www.statesacademicpress.com

STATES
ACADEMIC PRESS

Contents

Preface

The branch of physics which deals with the fundamental properties of sub-atomic particles that constitute radiation and matter is known as particle physics. A few of such particles are photons, neutrons, protons, electrons, muons, and neutrinos. They exhibit a wave particle duality and their dynamics are governed by the principles of quantum mechanics. Properties such as electrical charge, spin, mass and magnetism are studied by making use of quantum state vectors in a Hilbert space. Particle physics finds extensive applications in the fields of medicine, computer science, electronics and national security. Different approaches, evaluations, methodologies and advanced studies on particle physics have been included in this book. Some of the diverse topics covered herein address the varied branches that fall under this category. As this field is emerging at a rapid pace, the contents of this book will help the readers understand the modern concepts and applications of this subject.

A foreword of all Chapters of the book is provided below:

Chapter 1 - The branch of physics which deals with the study of the nature of particles which make up matter and radiation is termed as particle physics. The theory which describes three of the four fundamental forces in the universe is known as the standard model of particle physics. This is chapter will provide a brief introduction to the significant aspects of particle physics and standard model; **Chapter 2 -** Some of the major theories which are studied within the discipline of particle physics are gauge theories, Higgs boson theory, quantum field theory, string theory, effective field theory and supersymmetry. This chapter has been carefully written to provide an easy understanding of the varied facets of these theories of particle physics; **Chapter 3 -** The particles which are smaller than atoms are known as subatomic particles. These can be either elementary particles or composite particles. The subatomic particles which have an integer spin are known as bosons and those that have an off half-integer spin are known as fermions. This chapter discusses in detail these two types of subatomic particles; **Chapter 4 -** Quarks are found within protons and neutrons, and are considered to be the building blocks of matter. The particles which experience the strong nuclear force are called hadrons. This chapter closely examines the key concepts of quarks and hadrons as well as color charge to provide an extensive understanding of the subject; **Chapter 5 -** The standard model of particle physics is a theory which seeks to classify all the known elementary particles as well as describe the electromagnetic force, weak nuclear force and strong nuclear force. The topics elaborated in this chapter will help in gaining a better perspective about the standard models of particle physics as well as its mathematical formulation.

I would like to thank the entire editorial team who made sincere efforts for this book and my family who supported me in my efforts of working on this book. I take this opportunity to thank all those who have been a guiding force throughout my life.

Aidan Butler

1

Particle Physics: An Introduction

The branch of physics which deals with the study of the nature of particles which make up matter and radiation is termed as particle physics. The theory which describes three of the four fundamental forces in the universe is known as the standard model of particle physics. This is chapter will provide a brief introduction to the significant aspects of particle physics and standard model.

Particle physics is the study of the fundamental particulate constituents of nature. Our knowledge of these constituents is important to understand the laws that shape our universe, how they manifest their will, and why things are the way they are. Because of the scale at which its experiments are conducted and the diverse branches of physics and mathematics its theories draw from, particle physics has many applications – ranging from solid-state physics to medical diagnostics to distributed computing.

Fundamental, or elementary, particles are the smallest constituents of something and aren't made up of even smaller particles. Particle physics addresses both matter and force in terms of particles, so fundamental particles come in two types as well. Bosons are force-carriers: if a particle is acted upon by a force, then it can also be understood as the particle interacting with the particulate carriers of that force. For example: if the weak force is acting on a particle, then the particle is interacting with W± and Z bosons, which are carriers of the weak force. Fermions are matter-makers: they come together under the influence of others like them as well as bosons to make up matter (or antimatter).

However, a deeper classification of bosons and fermions alludes to their collective properties when they accumulate in large numbers. All particles have a spin quantum number associated with their angular momentum, expressed as integer fractions (0, 1/2, 1, 3/2 ...). All fermions have half-integer spins (1/2, 3/2 ...) and their behaviour is thus defined by the rules of Fermi-Dirac statistics. All bosons have integer spins (0, 1, 2 ...) and their behaviour is thus defined by the rules of Bose-Einstein statistics.

One commonly cited difference between the statistics is that bosons don't obey Pauli's exclusion principle while fermions do. The principle states that, given all fermions can be accorded four quantum numbers that define the particles' states, no two particles can have the same four numbers at the same time. And because bosons can possess such identical quantum numbers, they can form Bose-Einstein condensates.

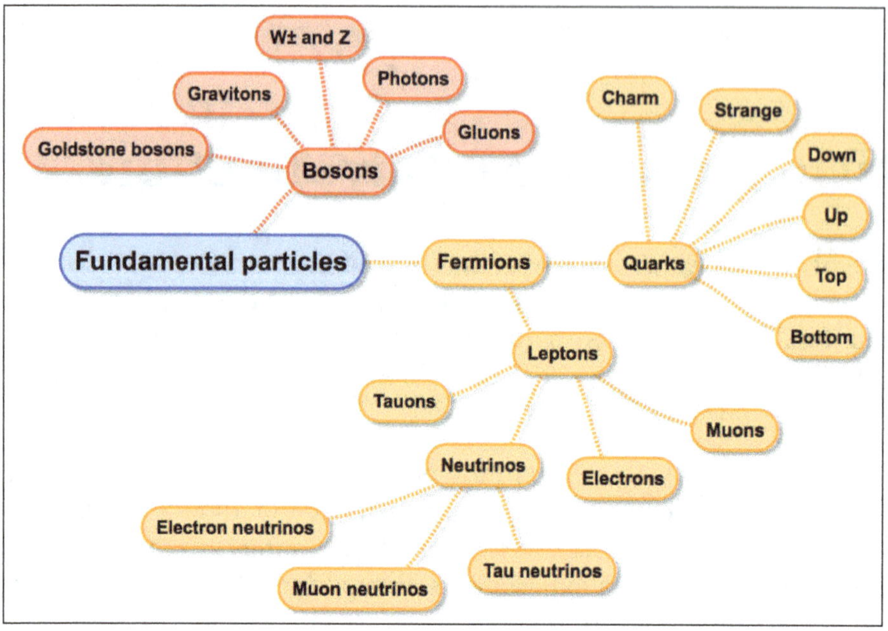

There are four forces that occur naturally. In the descending order of their strength, they are the strong nuclear force (holds nuclei together), the electromagnetic force (acts between charged particles), the weak nuclear force (responsible for radioactivity) and gravitation (acts between things having mass). They are respectively mediated by gluons, photons, W±/Z bosons and the hypothetical gravitons. At higher energies, the electromagnetic and weak nuclear forces unify into one common electroweak force. Some physicists think that at even higher energies, beyond those we can access with the current state of experimental physics, all four forces will unify to form one 'grand unified' force. The theories that aspire to account for such unification are called grand unified theories.

Accelerators are used to take advantage of three concepts that let us access the extremely small distances inhabited by particles.The first is the special theory of relativity, formulated by Albert Einstein in 1904. Its most famous implications are that mass and energy can be converted into each other, and that objects become more massive as they travel at speeds closer to that of light. As a result, when an accelerator accelerates particles, they become more energetic, and the excess energy is used to create new particles. What particles can or can't be created depends on the experimental conditions as well as what theoretical models allow. The output is also limited by the total energy available during the collision (13 TeV in the LHC's case).

The mass-energy equivalence dictates that the energy and mass of a particle are related by a factor of c^2, c being the speed of light. Specifically, $m = E/c^2$. So, when a physicist says a particle weighs 1 electron-volt, a unit of energy relevant to particles, she's actually saying it weighs 1 eV/c^2. Secondly: a common benchmark is the proton's rest mass, about 938 million eV = 0.938 GeV.

The second concept has to do with the principles of quantum mechanics, which allow heavier particles to break down into combinations of lighter particles in a process called decaying. How often they decay into which particles is dictated by parameters called coupling constants. Let's say a heavier *H* particle decays into two *P* particles some of the time, two *Q* particles at other times, and never into any *R* particles. If the *H* and *P* coupling constant is higher than the *H* and *Q* coupling constant, then *H* will decay into two *P*s proportionally more often than it will decay into two *Q*s. And the *H* and *R* coupling constant will be zero. Based on which particle physicists are trying to study, detectors stationed around the accelerator tunnel will be tuned to look for certain combinations of decay products.

The third concept involves the wave-function. It's a number that describes the probability of a particle existing at a particle point in space and time, and is calculated in terms of a wave – i.e. possessing a frequency and a wavelength. The wavelength corresponds to the distance across which the particle will manifest itself. Since energy and frequency are directly proportional, energy and wavelength are inversely proportional. The implication is that the higher the energy an accelerator is able to push particles to, the smaller the particles it will be capable of finding.

Use of Particle Accelerators

All particles behave differently when passing through matter. Electrons and photons rapidly lose their energy by continuously colliding with other material atoms until they get trapped. Protons travel through matter almost unnoticed until they are close to the end of their journey, when they lose all their energy in a burst and immediately stop. Muons, on the other hand, can travel through layers and layers of any material, even though hundreds of meters of solid rock, without losing much of their energy. Scientists build labs, like the Gran Sasso laboratory in Italy, deep under mountains to shield sensitive experiments from cosmic muons. Neutrinos, another fundamental particle, interact extremely weakly with all other objects: they can pass through entire planets without stopping.

Physicists leverage these different behaviors to develop detectors that catch and measure each type of particle. When a particle passes through a material it leaves a certain amount of its energy behind. That energy is collected and transferred to an electronic interface which visualizes a particle's passage. By layering different detectors, physicists can reconstruct particle trajectories, and, under certain circumstances, measure the exact energy of a particle.

Many particles have an electric charge. A charged particle behaves like the needle of a compass near a magnet: it points towards one of the magnet's poles. Charged particles, like electrons or protons, are influenced by electromagnetic fields. An electric field can be used to push charged particles forward, while a magnetic field can make a particle move to one side or another, depending on the charge of the particle.

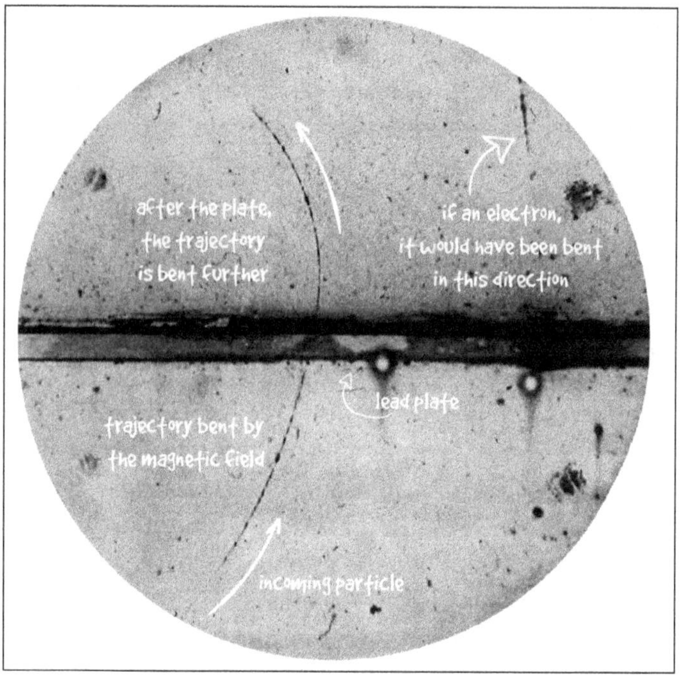

In 1933, while photographing cosmic rays with a Wilson chamber, Carl Anderson took a photograph of a particle passing through a thin film of photographic emulsion. The particle behaved almost exactly like an electron, except for the way a magnetic field around the film bent its trajectory. If it were an electron, the magnet would have bent the path in the opposite direction. Anderson had proof that this particle wasn't an electron. What he observed was a positron, the anti-electron, and the first proof of the existence of anti-matter.

Accelerating Particles

Electromagnetic fields influence the speed and trajectories of charged particles. An electromagnetic field can always be divided into two distinct components: an electric field and a magnetic field. The force exerted by the field on a charged particle is called the Lorentz force:

$$F = q\left[E + (v \times B)\right]$$

q is the charge of the particle and v its velocity, E is the electric field, and B is the magnetic field. For a given electric and magnetic field, the particle experiences a force F from the field dependent on its charge and speed. Rewriting the formula to isolate the velocity of the particle we find this:

$$v = \sqrt{\frac{2q(V_2 - V_1)}{m}}$$

The expression describes the velocity of a particle, v, given the potential difference of an electric field. This means that creating a large voltage difference will increase a charged particle's velocity.

The particle in the diagram has been drastically slowed down. In reality, the particle would instantly fly out of the page, since an electron put in a small accelerator (say, of 10 meters) and pushed by an electric field (say, of 100 Volts), will be accelerated to a speed of about 6 million meters per second—that is almost 22 million km/h.

That speed, while fast on its own, is still much slower than the speed of light—the fastest possible speed in nature, as far as we currently know—, which is about 1 billion km/h.

Also notice how much more energy is needed to push a proton, which has the same electric charge as an electron, but contains a much higher mass. You have to choose the highest voltage setting to make it move.

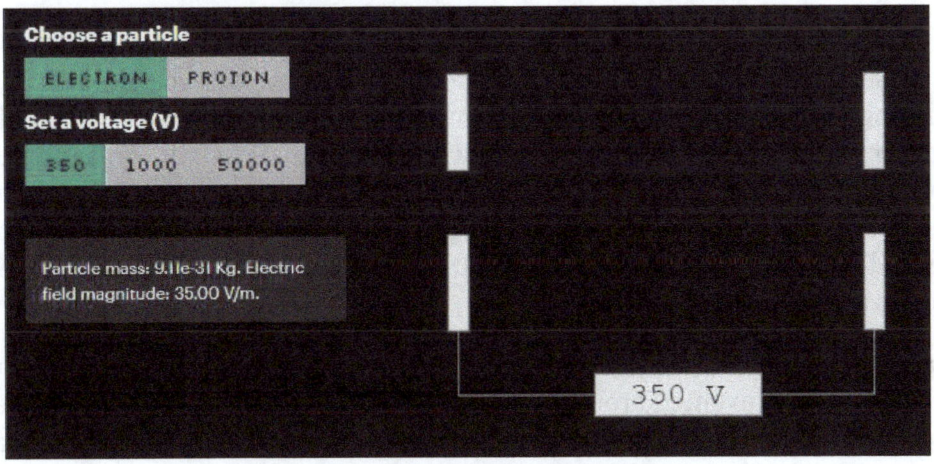

Accelerated a particle using an electric field.

If we instead rewriting the formula to isolate the radius of the bending of its trajectory, we find a description of the radius of the curvature of a particle's trajectory, r given the speed of the particle and the strength of the magnetic field:

$$r = \frac{mv}{qB}\sin\theta$$

Means that a magnetic field can be used to steer the path of a charged particle.

In reality, this process is much more complicated. Different types of accelerating modules are used to push different types of particles, and many types of magnets (e.g., dipoles, quadrupoles, sextuples, and so on) are used to bend, squeeze, and focus beams of charged particles to maximize the their intensity (as seen in the accelerators used for medicine or industry applications) or the number of their collisions (as seen in the accelerators used for physics experiments).

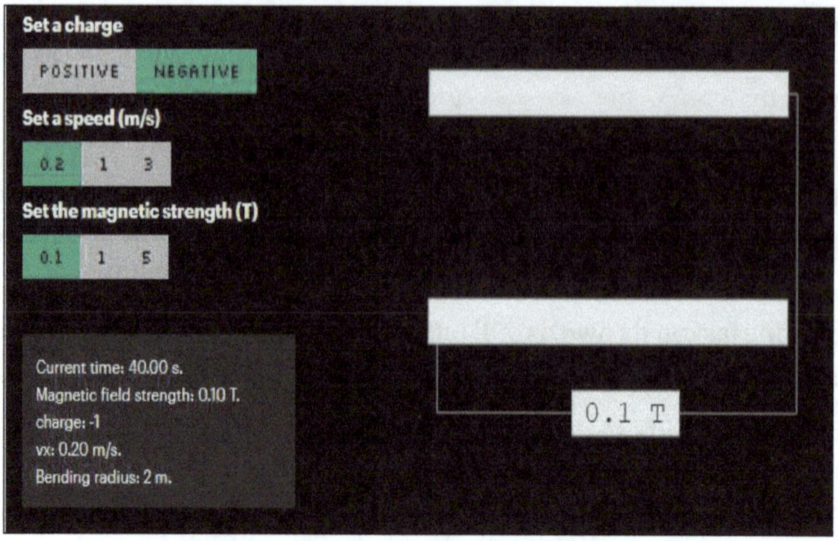

Steer a particle using a magnetic field.

Using these two principles, physicists started building particle accelerators. By using accelerating modules one after another, we can build linear accelerators where the energy that these accelerators can produce is governed by their length. By combining accelerating and bending modules, we can build circular accelerators, in which particles are injected and forced to move in a circular path. Each time a particle goes around it is accelerated. After a given amount of cycles the particles attains their maximum energy and can be used for experiments.

Circular accelerators can be further divided into two main groups: cyclotrons, which were the first to be invented around 1930, and synchrotrons. The former can be seen in hospitals today. The latter are used in machines for cutting-edge research in particle physics, such as the Large Hadron Collider (LHC) at CERN.

Physicists designed accelerators to study the basic characteristics of particles, but it quickly became clear that they could have many other scientific and industrial applications. For example, old television sets were driven by cathode-ray tubes, which are nothing more than compact particle accelerators.

Electrons are produced and subjected to an electric field, causing them to fly towards a certain end of the tube. This end is covered by a layer of phosphorescent material that becomes luminescent when hit by electrons. The trajectory of the electrons is bent by a variable magnetic field, which forces electrons to hit the luminescent screen at different points. By modulating the intensity of the magnetic field, the electron beam paints an image on the screen.

Particle Factories

There are many more natural phenomena involving particles. All fundamental phenomena involving the electromagnetic force, for example, depend on the characteristics of electrons and photons—the elementary particle which is responsible for light.

The stability of atoms, which compose all matter, depend on the interaction between two fundamental particles: quarks and gluons. These particles were discovered at the SLAC laboratory (US) at the end of the 1960s and at the DESY Laboratory (Germany) in 1979, respectively. The mechanism behind radioactive decay—which occurs in the cores of stars and planets like Earth—is determined by the behavior of the WW and ZZ particles, discovered at CERN in 1983.

Many fundamental mechanisms in the universe depend on particles which we haven't discovered yet. This creates a number of interesting questions for modern physicists. For example, how did the universe form? At the beginning of time, in a fraction of a second, quarks and gluons bounced and smashed, interacting and recombining until creating the first glimpse of the matter we see today. And which particle, if any, is behind Dark Matter, which composes about 25% of the total energy of our universe? Is gravity carried by a particle like the other fundamental forces? And how are high-energetic cosmic rays that hit our planet generated, if there are no known sources for them near Earth?

To answer these questions, we need particles interacting at much higher energies and with a much higher flux than those we can obtain from natural sources like radioactive elements or cosmic rays. To make progress, we would need bespoke particle factories, to create large fluxes of particles at the desired energy to study interactions and effects. That's where Einstein comes in.

The particles created in colliders are not to be considered artificial: they are naturally found under specific conditions in other places, like stars, or confined in microscopic objects, like the nuclei of atoms. Using particle colliders, we are simply creating the right conditions for them to be observed in a laboratory. Although today's particle accelerators are impressive, they don't compare to natural particle accelerators found in outer space. Cosmic rays from outside the Milky Way create much more energetic collisions in the Earth's upper atmosphere than our most powerful colliders.

Conducting Experiments on Particles

In a circular collider, two beams of accelerated particles smash into one another at a collision point. From the energy of that collision, new particles arise, which physicists can catch and measure. By identifying and measuring these particles, physicists can infer how they were generated. For that, particle physics experiments use many layers of different types of detectors placed around the collision point, since different particles require different detectors (different particles act differently when traveling through matter).

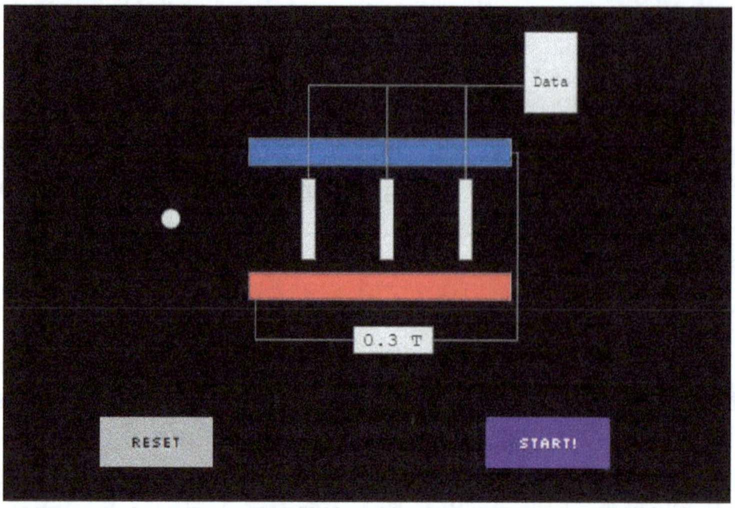

Detect a particle and measure its trajectory: By measuring the curve taken by the charged particle in the magnetic field we can infer its energy.

In the simplified visualization above, three layers of detectors are placed one after the other. A magnetic field permeates the space around them. When a charged particle comes, the magnetic field bends its trajectory, and makes it traverse the three detectors at different position. The detectors record the passage of the particle and send the information to the computers which handle the experimental data. From those data, scientists can reconstruct the trajectory of the particle in the experiment and, by knowing the strength of the magnetic field, the energy of the particle.

In the simplified visualization above, three layers of detectors are placed one after the other. A magnetic field permeates the space around them. When a charged particle comes, the magnetic field bends its trajectory, and makes it traverse the three detectors at different position. The detectors record the passage of the particle and send the information to the computers which handle the experimental data. From those data, scientists can reconstruct the trajectory of the particle in the experiment and, by knowing the strength of the magnetic field, the energy of the particle.

Modern particle physics experiments can be very large due to the amount of energy involved smashing particles together. When a collision occurs, many new particles are created which move outward from the collision at extremely fast speeds in all directions. To catch and measure these particles properly, scientists use layers of detectors and magnets placed around the collision point, so that they can reconstruct the particle trajectory after the collision for analysis.

In reality, many collisions occur and generate particles. The LHC, for example, smashes particles every 25 nanoseconds, producing about 40 million collisions every second! In each collision, thousands of particles are produced stemming from many different underlying phenomena. The work of experimental particle physicists concerns the design and development of bespoke hardware and software systems to effectively acquire and filter data from these collisions. With this data, they can identify and measure characteristics of particles and ultimately learn about the underlying mechanisms which generate them.

Particle Physics in Society

Particle accelerators such as those at CERN and Fermilab are designed and built for cutting-edge scientific research, but there are many others which are used for very different, practical applications. Without noticing, you may have already interacted with particles accelerators and particle detectors. For example, if you have ever had a dental X-ray, or a PET scan, then you have experienced a compact version of a particle accelerator and detectors.

There are about 30,000 particle accelerators in the world today, almost half of which are used in medicine in applications such as radiotherapy, X-ray imaging, and particle therapy. Some are used for other purposes, such as producing semiconductors. In fact, only 1% of all particle accelerators running today are used for fundamental research.

Many technologies we commonly use today are the results of research spin-off from particle physics. These byproducts were developed to build and run experiments or to support the activities of the particle physics community. Two impactful examples include medical scanners used in hospitals and the World Wide Web. Both are indirect products of fundamental research conducted by particle physicists while conducting their primary research.

Medicine

Particle physics and medicine have a long history, dating back to some of the first particle physics experiments every conducted: observing particles that interact with our bodies. Thousands of low energy particles hit our skin every second without us noticing. Our skin acts as a barrier and is strong enough to stop particles, ultimately protecting our internal organs. But more energetic particles can pierce through our bodies. This was observed by early experimenters who saw the effects of energetic particles traveling through their hands and bodies while setting up experimental apparatuses. This discovery spawned hundreds of new experiments within medicine, whose results are still used today for many diagnosis techniques, such as radiographies and CT scans.

What these early pioneers did not know however, is that energetic particles release their energy during their path through our bodies, and produce different biological effects in the tissues and organs they traverse. If not properly controlled, energetic particles traveling through an organ can be very dangerous. But, if controlled and well-calibrated, these techniques can visualize sick organs or heal malignant tumors where traditional surgery cannot reach. Today, medical physics is an independent branch of research, which studies, "the application of physics techniques to the human health".

X-Ray Images

Many people get a radiography—that black and white image of bones or other organs—at least once in their lives, perhaps to confirm a broken bone or to inspect dental health. Radiography imaging, better known as X-rays, is a direct application of what was discovered from early experiments in particle physics.

X-rays (where the "X" stands for "unknown", originally named for its mysterious origins) were found in 1895 by German physicist Wilhelm Röntgen. He discovered them while experimenting with a Crookes tube: the forerunner of the cathode-ray tube used in TV sets. Röntgen noticed a luminescent halo in the fluorescent screen far away from the tube, even when thick books and a thin metal plate were placed between them. Röntgen realized that those rays (which have been later discovered to be photons of a certain energy) were absorbed differently by different materials. By placing a photographic plate behind the object to test, he could observe the level of darkness in the resulting image depending on the amount of energy that particles left in the object. Shortly after, Röntgen had a "Eureka!" moment when he realized the new "X"-rays could also pass through the human body, producing an image of what's inside. He then took the first X-ray image, using his wife's hand and her wedding ring as the subject.

PET Scanners

Besides X-rays, there are many other medical imaging techniques that apply physics techniques. Positron-emission tomography (PET), for example, is a nuclear medicine imaging technique, which comes directly from particle and nuclear physics research.

PET scanners use a radioactive isotope, called "tracer", which binds to specific tissues in the patient's body. The isotope then rapidly decays emitting positrons—positive electrons. The positron travels a very short distance in the body before releasing its energy, and is eventually caught by an atom, where it meets an electron. When a positron, the anti-matter buddy of the electron, meets an electron, the two particles annihilate. That means that they "dissolve" during the interaction and a certain amount of energy is released, following Einstein's mass-energy equivalence. The annihilation energy produces two photons of a given energy, which escape the patient's body in opposite directions.

To catch the escaping photons, layers of particle detectors are placed around the patient—in the very same way particle detectors are placed around the collision point when studying the interaction between fundamental particles. The detectors around the patient—the big ring patients see when they lay down for the PET exam—detect the two escaping photons, measuring the time of arrival and the direction. A piece of dedicated software reconstructs the trajectory and the point of origin within the patient's body.

The tracer collects in organs and tissues of higher chemical activity, often bound to certain disease. The areas where the tracer accumulates are the origin of a higher number of photons and show up as bright spots on the images from the PET scan. Cancer cells, for instance, have a higher metabolic rate than normal cells and, because of this, they appear as bright spots on a PET scan, helping doctors check the presence or the status of a tumor.

The radioactive isotopes used in PET scanners have to be produced in cyclotrons, small circular particle accelerators which combine an oscillating electric field and a static magnetic field that holds the particles to accelerate in a spiral path. The accelerated particles are then driven to a specific target which, after irradiation, contains the desired radioisotope. The parameters like particle energy, beam intensity, target material, and time will determine the kind and the quantity of produced radioisotopes.

Decay after decay, samples of radioactive materials slowly "die", emitting fewer and fewer particles. How slow they die it depends on the material itself. The half-life is a parameter that indicates the time needed for half of the atoms of the radioactive material to decay, on average; that, in practice, indicates the duration of the radioactive material in time. Tracers with a shorter half-life must be used within a short delay from their production, therefore, they should be produced closed to the PET facility or to be transported there as fast as possible. PET radio-isotopes can be transported to maximum 3-5 hours distance only, which confines their production within the hospitals neighborhood.

Since the half-life of the produced radioactive tracers is quite short—short-lived tracers are chosen specifically to keep the radioactive dose absorbed by the patient as low as possible—, those have to be produced close to the PET scanner. One of the most used PET tracers is based on the radioisotope fluorine-18, whose half-life is of 110 minutes, which allows for medium-long transports. Another one is rubidium-82, with a half-life of 1.27 minutes, which is usually produced very close to the PET facility.

Given that the cost of operating a cyclotron, along with the equipment for the subsequent preparation of the radioactive tracer, are quite high, those are usually found only in large hospitals or in universities and research institutes which operate those facilities for different smaller hospitals simultaneously.

New Frontiers of Cancer Therapy: Hadron Therapy

Particles are used as therapies too. Beams of ionizing radiation—another name to indicate beams of energetic particles damage the DNA of cancerous tissue leading to their cellular death.

In 1906, shortly after the discovery of radioactivity itself by Henri Bequerel, Pierre Curie suggested that tumors could be cured by inserting a radioactive source—usually emitting electrons or photons—, which causes them to shrink. The technique was called brachitherapy, from the Greek word brachys, which means "short distance", to indicate that the radioactive source is placed in contact with the tumor. Common in the beginning of the 20th century, brachitherapy is still used today to treat different kind of tumors.

Another common cancer therapy which makes use of particles is radiotherapy, where X-rays are collimated and driven to the target organ or tissue. To deliver the planned dose, multiple treatments from different angles are usually given. Electron therapy uses beams of energetic electrons. Given their characteristics, beams of electrons do not penetrate deeply in the body tissues, so they are mainly used to treat skin tumors.

The drawback of all these techniques resides in the nature of the particles used for them. Because of the way energetic electrons or photons interact with matter (all matter, not just bodily tissue), much energy is released in the healthy tissue along a particle's path through a patient's body. This causes damage to the other tissue and organs near the tumor. Even with careful planning of the treatment, some energy will be released in the heathy organs around the sick ones, causing them to get damaged; and that is due to the very nature of the particles used for the treatment.

Protons, neutrons, and ions all belong to the same family of hadrons. Those are the particles composed of the smaller, fundamental bricks called quarks. Therefore, the term hadron therapy is used for all therapies making use of hadrons.

A new frontier in cancer therapy with particle beams is hadron therapy, which makes use of protons or heavier ions—that is, nuclei of heavier elements, like carbon. Because of the way hadrons interact with matter, they release the most significant part of their energy only at the end of their travel within a particular target; and the length of that path in the target depends on the initial beam energy. Thus, by fine-tuning the energy of the protons, we can make them hit almost only the tumor tissues, make them kill only the cancerous cells. Like a very precise "proton blade".

Standard Model of Physics

The Standard Model is the name given in the 1970s to a theory of fundamental particles and how they interact. It incorporated all that was known about subatomic particles at the time and predicted the existence of additional particles as well. There are seventeen named particles in the Standard Model, organized into the chart shown below. The last particles discovered were the W and Z bosons in 1983, the top quark in 1995, the tau neutrino in 2000, and the Higgs boson in 2012.

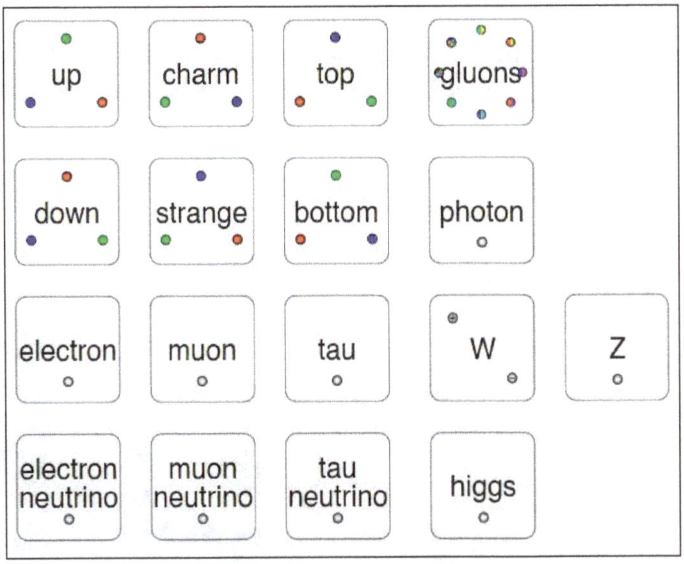

The fundamental particles of the standard model.

family			p particle	predicted/ discovered		spin number	charge (e)	color	*mass* (MeV/c²)
fermions	quarks	u	up quark	1964	1968	½	+⅔+	r, g, b	2.16
		d	down quark	1964	1968	½	−⅓−	r, g, b	4.67
		c	charm quark	1970	1974	½	+⅔+	r, g, b	1,270
		s	strange quark	1964	1968	½	−⅓−	r, g, b	93
		t	top quark	1973	1995	½	+⅔+	r, g, b	172,760
		b	bottom quark	1973	1977	½	−⅓−	r, g, b	4,180
	leptons	e	electron	1874	1897	½	−1−	none	0.51099895
		μ	muon	0000	1936	½	−1−	none	105.658375
		τ	tau	0000	1975	½	−1−	none	1776.86
		ν_e	electron neutrino	1930	1956	½	0	none	$< 1.1 \times 10^{-6}$
		ν_μ	muon neutrino	s1940s	1962	½	0	none	< 0.19
		ν_τ	tau neutrino	s1970s	2000	½	0	none	< 18.2
	†	p	proton	1815	1917	½	+1+	none	938.272081
		n	neutron	1920	1932	½	0	none	939.565413
bosons	vector	g	gluon	1962	1978	1	0	8 colors	0
		γ	photon	0000	1899	1	0	none	0
		W	W boson	1968	1983	1	±1±	none	80,379
		Z	Z boson	1968	1983	1	0	none	91,187.6
	‡	H	higgs boson	1964	2012	0	0	none	125,100

Standard model particles (nucleons included for comparison).

* Masses from Particle Data Group. † The proton and neutron are hadrons (made of quarks), baryons (made of three quarks), and nucleons (found in the nucleus), but they are not standard model particles.‡ The higgs boson is the only known scalar boson.

Particle Families

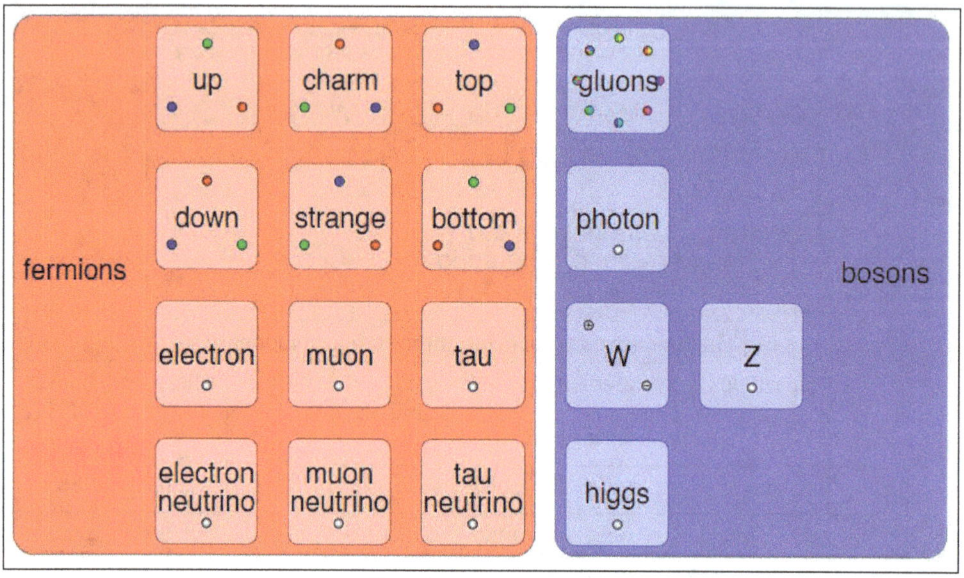

Fermions vs. Bosons.

Fundamental particles are either the building blocks of matter, called fermions, or the mediators of interactions, called bosons. There are twelve named fermions and five named bosons in the Standard Model.

Fermions obey a statistical rule described by Enrico Fermi of Italy, Paul Dirac of England, and Wolfgang Pauli of Austria called the exclusion principle. Simply stated, fermions cannot occupy the same place at the same time. (More formally, no two fermions may be described by the same quantum numbers.) Leptons and quarks are fermions, but so are things made from them like protons, neutrons, atoms, molecules, people, and walls. This agrees with our macroscopic observations of matter in everyday life. People cannot walk through walls unless the wall gets out of the way.

Bosons, in contrast, are have no problem occupying the same place at the same time. (More formally, two or more bosons may be described by the same quantum numbers.) The statistical rules that bosons obey were first described by Satyendra Bose of India and Albert Einstein of Germany. Gluons, photons, and the W, Z and Higgs are all bosons. As the particles that make up light and other forms of electromagnetic radiation, photons are the bosons we have the most direct experience with. In our everyday experience, we never see beams of light crash into one another. Photons are like phantoms. They pass through one another with no effect.

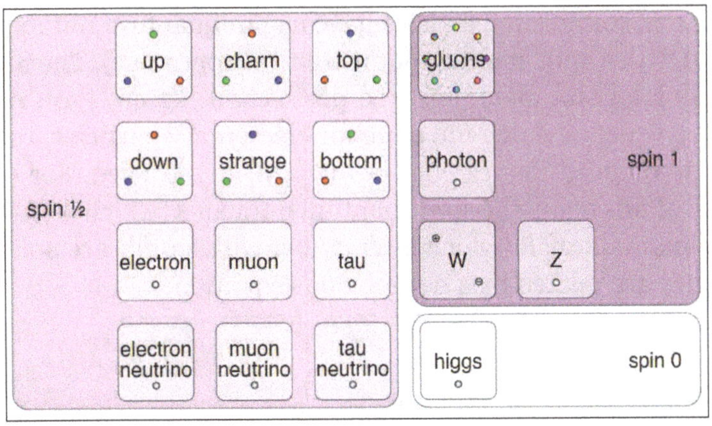

Elementary particles have an intrinsic spin angular momentum S. The adjective intrinsic means innate or essential to the thing itself. Elementary particles don't have spin because someone is spinning them. They just spin — or rather, they just have a measurable quantity with the same units as angular momentum. In current physics, elementary particles are featureless — like a mathematical point. In order for something to be perceived as spinning, the thing spinning would need something like a "front" and a "back". Featureless, point particles don't have anything like that. Particle physics is best described with mathematics. Spin is a convenient label for a measurable quality and not a description of reality.

Every elementary particle has associated with it a spin quantum number s (often called the spin number or just the spin), where s is any whole number multiple of a half. Fermions have half integral spin quantum numbers (½, 1½, 2½, etc.) and bosons have integral spin quantum numbers (0, 1, 2, etc.). No spin numbers are possible in between these. Spin is a quantized quantity.

The elementary fermions have a spin of ½. Particles made from combinations of fermions will have an overall spin that's a combination of the individual spins. A baryon composed three quarks will combine to an overall spin of ½ or 1½, since those are the only possible, non-negative combinations of ½ ± ½ ± ½. That shows that all baryons (like protons and neutrons, for example) are also fermions. Likewise, a meson composed of a quark and an antiquark will combine to an overall spin of 0 or 1 since those are the only possible, non-negative combinations of ½ ± ½. That shows that all all mesons (like the pion of the residual strong interaction, for example) are also bosons.

The force carrying bosons of the Standard Model (gluons, photons, and the W and Z) have spin 1 since they go with vector fields. The Higgs boson corresponds to a scalar field so it has spin 0. If the particle of the gravitational field is ever discovered, it would be called a graviton and would have a spin 2 since it corresponds to a tensor field. A tensor is a mathematical object that's more complex than a vector, which is in turn more complex than a scalar. See the trend? A scalar field with no direction gets a particle with spin 0. A vector field with a direction gets a particle with spin 1. A tensor field that stretches and squeezes space in two directions gets a particle with spin 2.

All fundamental and composite particles have a spin quantum number s (lowercase). This is associated with a spin angular momentum S (uppercase). The SI unit of angular momentum is the kilogram meter squared per second [kg m^2/s] or, equivalently, the joule second [J s], which is much too large for elementary particles. Instead $\hbar$ (h bar), also known as the reduced Planck constant ($\hbar = h/2\pi$), is used. For reasons that are beyond the scope of this book, the spin quantum number s (which is just a number) and the spin angular momentum S (which is a number with a unit) are not numerically the same. Instead, they are related by a non-obvious equation:

$$S=\left[s\left(s+1\right)\right]^{1\backslash 2}\hbar$$

For particles with spin quantum number of 0, the solution is a sensible spin angular momentum of 0 $\hbar$:

$$S(0)=\left[0\left(0+1\right)\right]^{1/2}\hbar=0\hbar$$

For higher spin quantum numbers the spin angular momentum increases, but beyond that there's not much else that can be simply said:

$$S\left(\tfrac{1}{2}\right)=\left[\frac{1}{2}\left(\frac{1}{2}+1\right)\right]^{1/2}\hbar=\frac{\sqrt{3}}{2}\hbar$$

$$S(1)=\left[1\left(1+1\right)\right]^{1/2}\hbar=\sqrt{2}\,\hbar$$

$$S\left(1\tfrac{1}{2}\right)=\left[\frac{3}{2}\left(\frac{3}{2}+1\right)\right]^{1/2}\hbar=\frac{\sqrt{15}}{2}\hbar$$

$$S(2)=\left[2\left(2+1\right)\right]^{1/2}\hbar=\sqrt{6}\,\hbar$$

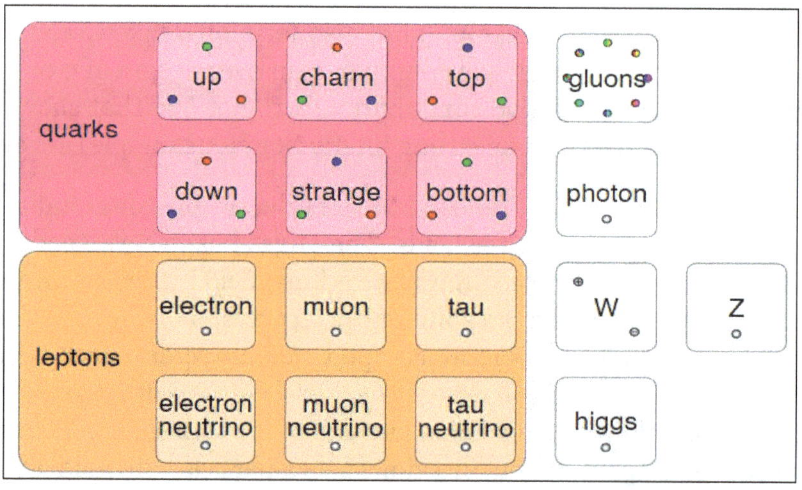

Fermions are divided into two groups of six: those that must bind together are called quarks and those that can exist independently are called leptons.

Quarks are known to bind into triplets and doublets. The triplets are called baryons, a term derived from the Greek word βαρύς (varys) meaning "heavy". The doublets are called mesons, a term derived from the Greek word μέσος (mesos) meaning "medium". Collectively baryons (the heavy triplets), mesons (the middleweight doublets), and quarks (the fundamental particles) are known as hadrons, from the Greek word αδρός (adros) meaning thick, robust, massive, or large. This name alludes to the ability of the point-like quarks to bind together and form particles that are "thick" in a certain sense.

The other six fermions are called leptons, a name derived from the Greek word λεπτός (leptos) meaning thin, delicate, lightweight, or small. These particles don't need to bind to each other, which keeps them "thin" in a certain sense. Originally leptons were considered the "light" particles and hadrons the "heavy" particles, but the discovery of the tau lepton in 1975 broke that rule. The tau (the heaviest lepton) is almost twice as massive as a proton (the lightest hadron). Baryons found in the nucleus (the proton and neutron) are called nucleons. The Latin word for kernel is nucleus. Nucleons are found in the metaphorical "kernel" of the atom. Baryons that contain at least one strange quark but no charm, bottom, or top quarks are called hyperons. The Greek word for beyond is υπέρ (yper), which morphed into the English prefix hyper-. Hyperons are particles that are "way out" in a certain sense.

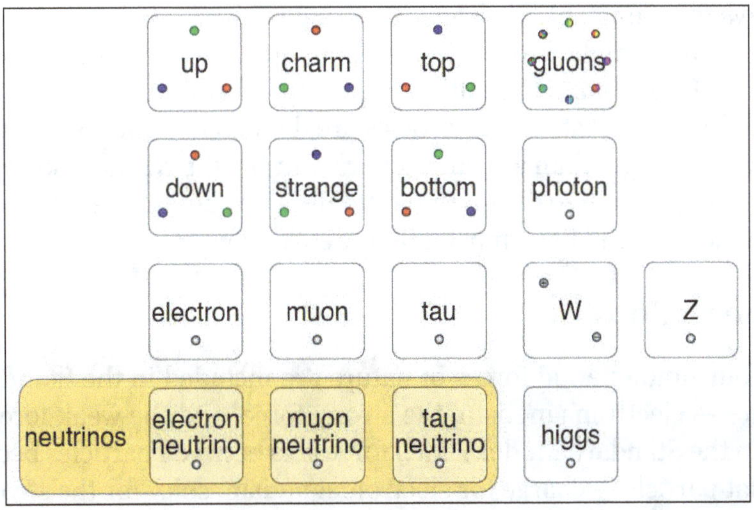

The neutrinos are an important subgroup within the leptons. They come in three flavors named for their partner leptons. The electron, muon, and tau are matched with the electron neutrino, muon neutrino, and tau neutrino. Neutrinos have very little mass (even for leptons) and interact so weakly with the rest of the particles that they are exceptionally difficult to detect. The name is a play on words. The Italian word for neutron (neutrone) sounds like the word neutral (neutro) with an augmentative suffix (-one) tacked on the end. Replace the augmentative suffix -one with the diminutive suffix -ino and you have a "little neutral", which is a good description of what a neutrino is — a diminutive neutral particle, so small and neutral.

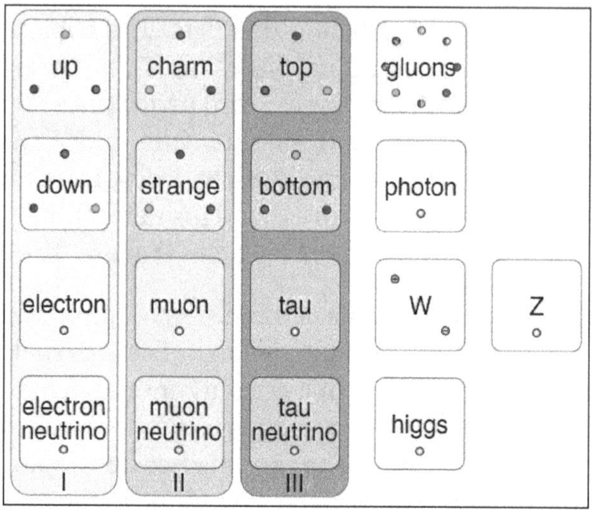

Fermion Generations.

Fermions belong to one of three known generations from ordinary (I), to exotic (II), to very exotic (III). (These are the adjectives I selected to describe the generations.) Generation I particles can combine to form hadrons with effectively infinite life spans (stable atoms made of electrons, protons, and neutrons for example). Generation II particles always form unstable hadrons. The longest lived hadron containing a generation II quark is the lambda particle (made of an up, down, and strange quark). It has a mean lifetime of less than a billionth of a second, which is considered long-lived for an unstable hadron. Generation III particles are divided in their behavior. The bottom quark isn't much stranger than a strange quark, but the top quark is so short-lived that it doesn't exist long enough to do anything. It falls apart before the world even knows it exists. Top quarks are only known from their decay products.

Particle Interactions

Three of the four fundamental forces of nature are included in the Standard Model of particle physics — electromagnetism, the strong force, and the weak force. (Gravity is not included in the Standard Model.) Each force acts between particles because of some property of that particle — charge for electromagnetism, color for the strong force, and flavor for the weak force. The bosons associated with each force are called gauge bosons — the photon for electromagnetism, gluons for the strong force, and the W and Z bosons for the weak force.

Charge is the property of matter that gives rise to electric and magnetic phenomena (known collectively as electromagnetism). Charge is quantized, which means it can only exist in discrete amounts with restricted values — multiples and fractions of the elementary charge ($e = 1.6 \times 10^{-19}$ C).

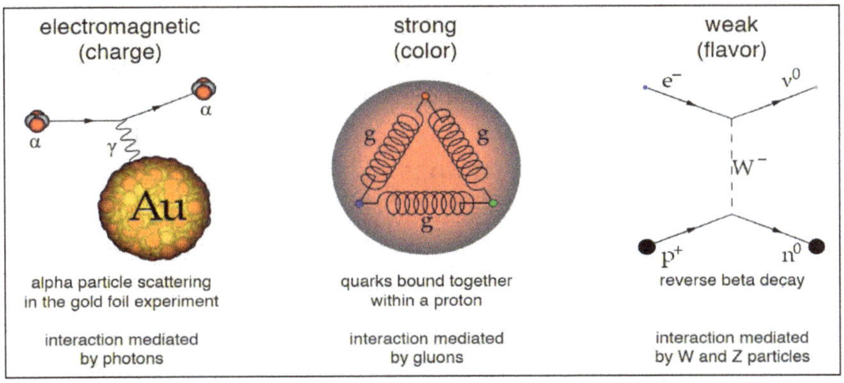

Particles that exist independently (the electron, muon, and tau) carry multiples of the elementary charge (−1 e), while quarks carry fractions of the elementary charge (+⅔ e or −⅓ e). Quarks always bind together in groups whose total charge is an integral multiple of the elementary charge, which is why no one has ever directly measured a fractional charge. In addition, since opposite charges attract, electrons tend to bind to protons to form atoms that are neutral overall. We don't normally notice the electrical nature of matter because of this.

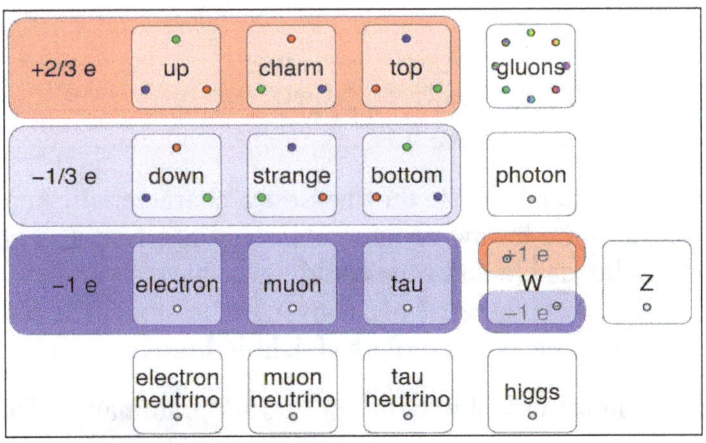

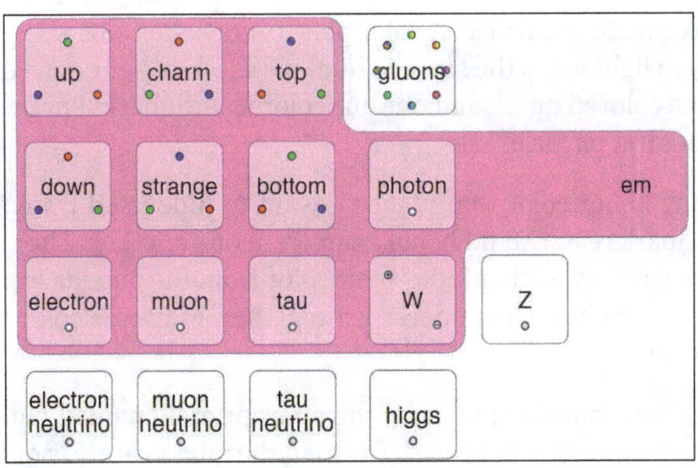

Charged particles interact by the exchange of photons — the carrier of the electromagnetic force. Whenever one electron repels another or any electron orbits a nucleus, a photon is responsible. Photons are massless, uncharged, and have an unlimited range. The mathematical model used to describe the interaction of charged particles through the exchange of photons is known as quantum electrodynamics (QED).

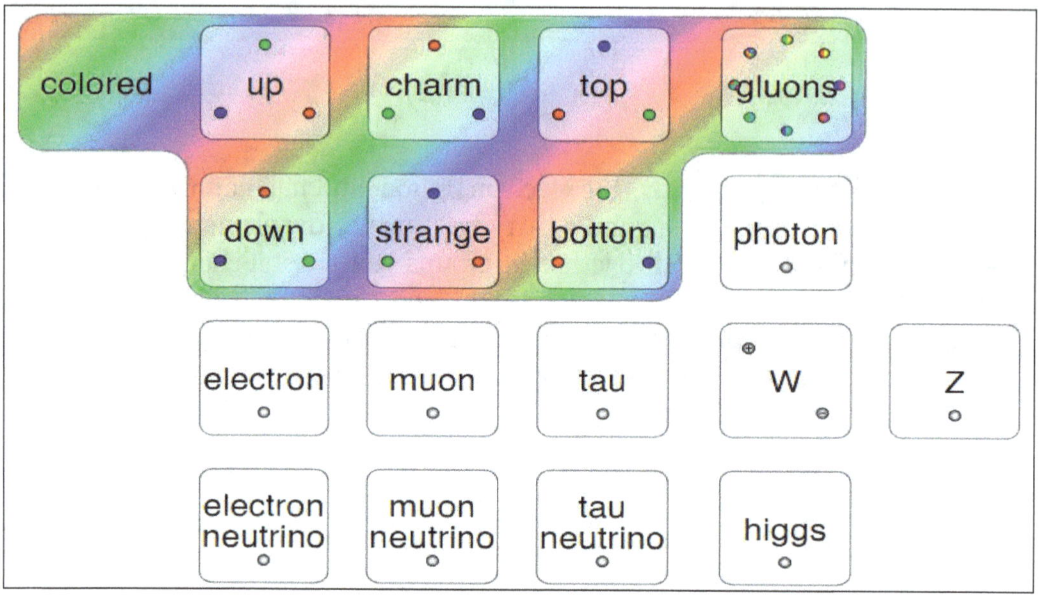

Quarks stick to other quarks because they possess a characteristic known as color (or color charge). Quarks come in one of three colors: red, green, and blue. Quarks are much too small to to be visible and thus could never have a perceptual property like color. The names were chosen because of a convenient analogy. The colors of quarks in the Standard Model combine like the colors of light in human vision.

Red light plus green light plus blue light appears to us humans as "colorless" white light. A baryon is a triplet of one red, one green, and one blue quark. Put them together and you get a color neutral particle. A color plus its opposite color also gives white light. Red light plus cyan light looks the same to humans as white light, for example. A meson is a doublet of one colored quark and one anticolored antiquark. Put them together and you get a color neutral particle.

There's something about color that makes it want to hide itself from anything bigger than a nucleus. Quarks can't stand being apart from one another. They just have to join up and always do so in a way that hides their color from the outside world. One color is never favored over another when quarks get together. Matter is color neutral down to the very small scale.

Colored particles are bound together by the appropriately named gluons. Gluons are also colored, but in a more complicated way than the quarks are. Six of the eight gluons have two colors, one has four, and another has six. Gluons glue quarks together, but they also stick to themselves. One consequence of this is that they they can't reach out and do much beyond the nucleus.

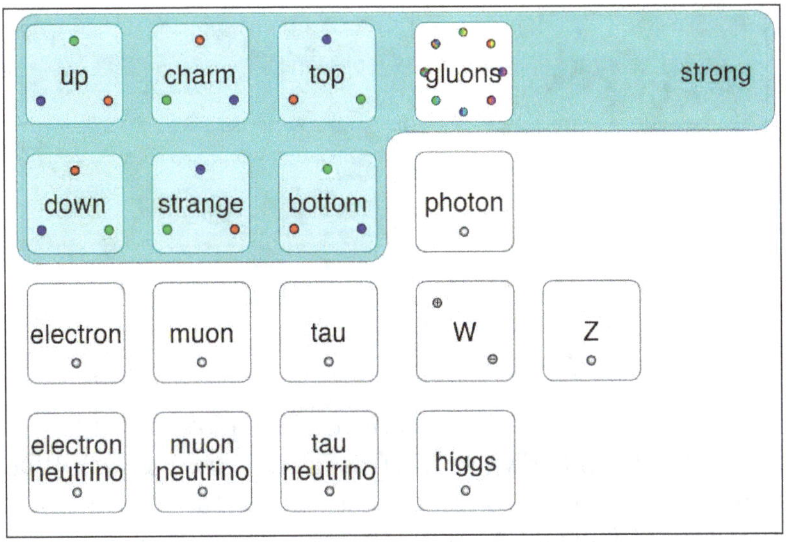

The mathematical model used to describe the interaction of colored particles through the exchange of gluons is known as quantum chromodynamics (QCD). The whole sticky mess is called the strong force or the strong interaction since it results in forces in the nucleus that are stronger than the electromagnetic force. Without the strong force, every nucleus would blow itself to smithereens.

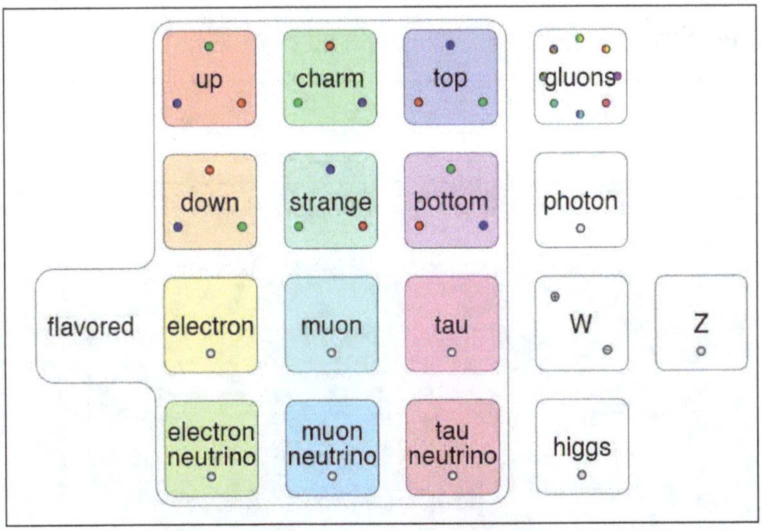

There are twelve named elementary fermions. The difference between them is one of flavor. The word "flavor" is used here to mean "type" and it applies only to fermions. Don't let the word mislead you. Subatomic particles are much too small to have any characteristics that could be directly observed by human senses.

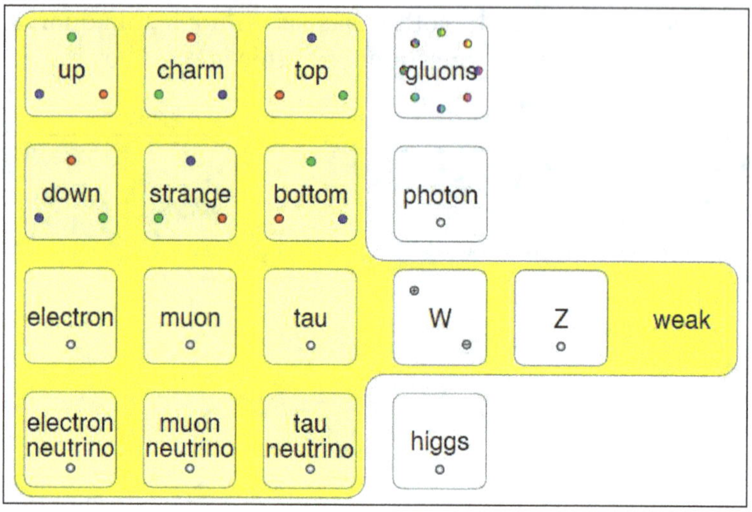

Flavored particles interact weakly through the exchange of W or Z bosons — the carriers of the weak force (also known as intermediate vector bosons). When a neutron decays into a proton, a W− boson is responsible. The mathematical model used to describe the interaction of flavored particles through the exchange of W and Z bosons is sometimes known as quantum flavor dynamics (QFD), but this is a term that is not used by working particle physicists. At higher energies, the weak and electromagnetic forces begin to look more and more alike. The mathematical model that describes these interactions together is known as electroweak theory (EWT). This is the conventional name for the theory of the weak force.

Mass and Gravity

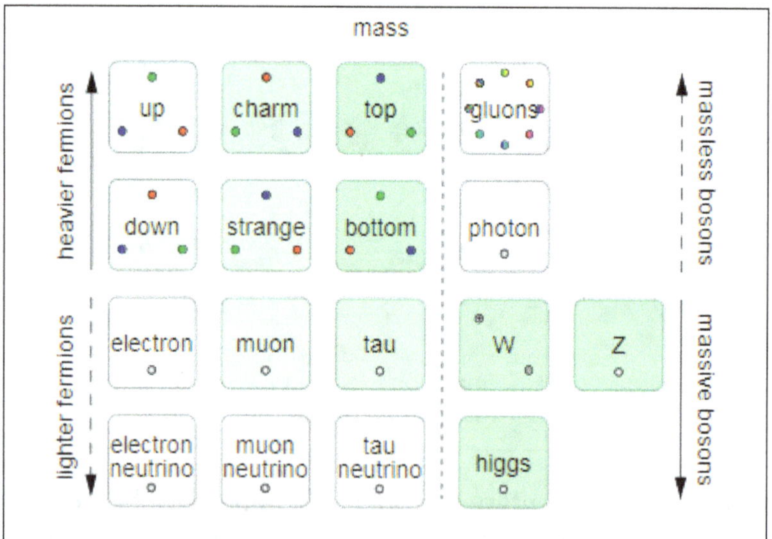

This chart shows general trends in mass.
(exception: $m_{up} < m_{down}$)

All fermions are thought to have a nonzero rest mass. Particles in generation I are less massive than those in generation II, which are less massive than those in generation III. Within the generations, quarks are more massive than leptons and neutrinos are less massive than the other leptons. Bosons are divided when it comes to mass. Gluons and photons are massless. The W, Z, and Higgs bosons are massive.

Mass is energy. A moving particle is more massive than a stationary particle because it has kinetic energy. Logically then, a stationary particle should have no mass. If we could stop a photon (which we can't) it would weigh nothing. If we could stop an electron (which we can) it would weigh something. Why do some particles weigh something at rest and others weigh nothing?

Mass is energy and energy comes in two types: kinetic energy (the energy of motion) and potential energy (the energy of arrangement). The kinetic energy contribution to mass is minor. Most of the mass around us comes from some sort of potential energy. For example, a proton is made of two up quarks and a down quark. The masses of these three quarks do not add up to the mass of a proton:

$$m_p \neq 2m_u + 1m_d$$

$$938.272\,\text{MeV}/c^2 \neq 2\left(2.3\ \text{MeV}/c^2\right) + 1\left(4.8\ \text{MeV}/c^2\right)$$

$$938.272\,\text{MeV}/c^2 \neq 9.4\,\text{MeV}/c^2$$

The masses of the parts are only 1% of the mass of the whole. The remaining 99% comes from the potential energy of the strong force holding the proton together. The particles that mediate the strong force are gluons. The interaction energy of these massless particles is what gives the proton most of its mass.

So why do quarks have mass but the gluons don't? Or as the question was historically stated, why do the W and Z bosons have mass but the photon doesn't? Maybe there's another kind of potential energy. Maybe there's another interaction out there — an interaction that some particles feel and others don't. If there's an interaction, there's a particle — a particle that gives mass to other particles when they're just sitting around doing nothing.

All of space is assumed to be filled with a Higgs field — a background sea of virtual Higgs bosons that pop in and out of existence. The quarks, leptons, and W and Z bosons moving around through space interact with this field, which is why these particles have mass. The photons and gluons do not interact with the Higgs field, which is why these particles do not have mass. Even the Higgs boson itself interacts with the Higgs field. It gives itself mass! The Higgs boson is different from the other bosons (gluons, photons, W and Z bosons) in that the Higgs mechanism doesn't result in anything resembling a force (like the strong, electromagnetic, and weak forces). The Higgs field is a scalar field and the Higgs boson is a particle with spin zero.

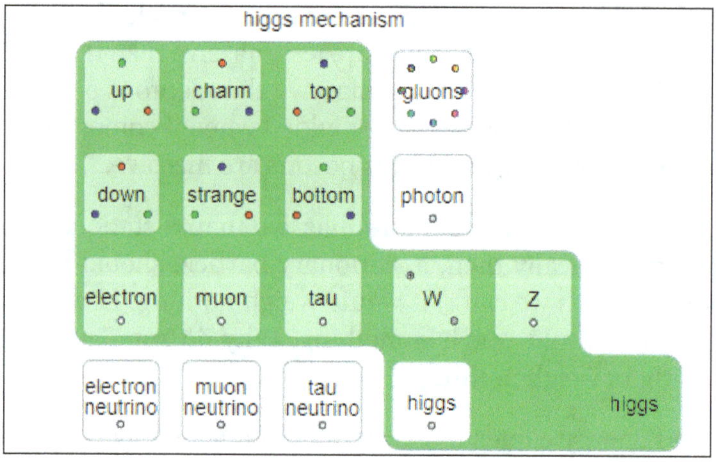

Interactions with the higgs field are mediated by the higgs boson, which is how most elementary particles get their mass.

Gravity is the force between objects due to their mass. The mathematical model that would describe gravity on the particle level is sometimes called quantum geometrodynamics (QGD), but is more often referred to as quantum gravitation. The Standard Model of particle physics does not include gravity (nor could it ever) and there currently is no quantum theory of gravitation. If there was, it would have to include a force carrying particle. The proposed name for this particle is the graviton. General relativity describes gravitational waves as a tensor disturbance that propagates — one that shears space-time along two alternating perpendicular directions. This two dimensional behavior leads theoretical physicists to believe that the graviton would have spin two.

References

- All-you-need-to-know-to-get-started-on-particle-physics: thewire.in, Retrieved 11, July 2020
- On-particle-physics: parametric.press, Retrieved 04, March 2020

2

Theories Related to Particle Physics

Some of the major theories which are studied within the discipline of particle physics are gauge theories, Higgs boson theory, quantum field theory, string theory, effective field theory and supersymmetry. This chapter has been carefully written to provide an easy understanding of the varied facets of these theories of particle physics.

Gauge Theories

Gauge theories refer to a quite general class of quantum field theories used for the description of elementary particles and their interactions. The theories are characterized by the presence of vector fields, and as such are a generalization of the older theory of Quantum Electrodynamics (QED) that is used to describe the electromagnetic interactions of charged elementary particles with spin 1/2. Local gauge invariance is a very central issue. An important feature is that these theories are often renormalizable when used in 3 space- and 1 time dimension.

Maxwell's Equations and Gauge Invariance

The simplest example of a gauge theory is electrodynamics, as described by the Maxwell equations. The electric field strength $\vec{E}\left(\vec{x},t\right)$ and the magnetic field strength $\vec{B}\left(\vec{x},t\right)$ obey the homogeneous Maxwell equations (in SI units):

$$\vec{\nabla}\times\vec{E}+\frac{\partial\vec{B}}{\partial t}=0$$

$$\vec{\nabla}\cdot\vec{B}=0$$

According to Poincaré's Lemma, $\vec{\nabla}\cdot\vec{B}=0$ implies that there exists another vector field $\vec{A}(\vec{x},t)$ such that:

$$\vec{B}=\vec{\nabla}\times\vec{A}$$

Since $\vec{\nabla}\times\vec{E}+\frac{\partial\vec{B}}{\partial t}=0$ now reads:

$$\vec{\nabla}\times\left(\vec{E}+\frac{\partial\vec{A}}{\partial t}\right)=0$$

We can also conclude that there is a potential field $\Phi(\vec{x},t)$ such that:

$$\vec{E}=-\vec{\nabla}\Phi-\frac{\partial\vec{A}}{\partial t}$$

The field Φ is the electric potential field; the vector field $\vec{A}$ is called the vector potential field. The strengths of these potential fields are determined by the inhomogeneous Maxwell equations, which are the equations that relate the strengths of the electromagnetic fields to the electric charges and currents that generate these fields. The use of potential fields often simplifies the problem of solving Maxwell's equations. What turns this theory into a gauge theory is the fact that the values of these potential fields are not completely determined by Maxwell's equations. Consider an electromagnetic field configuration $\left(\vec{E}(\vec{x},t),\vec{B}(\vec{x},t)\right)$ and suppose that it is described by the potential fields $\Phi(\vec{x},t),\vec{A}(\vec{x},t))$. Then, using any arbitrary scalar function $\Delta(\vec{x},t)$, one can find a different set of potential fields describing the same electric and magnetic fields, by writing,

$$\Phi'=\Phi+\frac{\partial\Lambda}{\partial t},$$

$$\vec{A}'=\vec{A}-\vec{\nabla}\Lambda$$

Inspecting equations, one easily observes that $\vec{E}=\vec{E}'$ and $\vec{B}=\vec{B}'$. Thus, the set $\left(\Phi',\vec{A}'\right)$ and $\left(\Phi,\vec{A}'\right)$ describe the same physical situation. Because of this, we call the transformation a gauge transformation. Since Λ may be chosen to be an arbitrary function of the points $(\vec{x},t)$ in space-time, we speak of a local gauge transformation. The fact that the electromagnetic fields are invariant under these local gauge transformations turns Maxwell's theory into a gauge theory.

In relativistic quantum field theory, the field $\psi(\vec{x},t)$ of a non-interacting spin less particle would typically obey the equation:

$$\left(\vec{\nabla}^2-\frac{\partial^2}{\partial t^2}\right)\psi=m^2\psi$$

Where units where used such that the velocity of light c=1, and Planck's constant $\hbar$=1. This gives the dispersion relation between energy and momentum as dictated by Special

Relativity:

$$E = \sqrt{\vec{p}^{\,2} + m^2},$$

Suppose now that the particle in question carries an electric charge q. How is its equation then affected by the presence of electro-magnetic fields? It turns out that one cannot write the correct equations using the fields $\vec{E}$ and $\vec{B}$ directly. Here, one can only choose to add terms depending on the (vector) potential fields instead:

$$\left(\vec{\nabla} - iq\vec{A}\right)^2 \psi - \left(\frac{\partial}{\partial t} + iq\Phi\right)^2 \psi = m^2\psi$$

This equation correctly produces waves that are deflected by the electro-magnetic forces in the way one expects. For instance, the energy E is easily seen to be enhanced by an amount $q\Phi(\vec{x},t)$, which is the potential energy of a charged particle in an electric potential field.

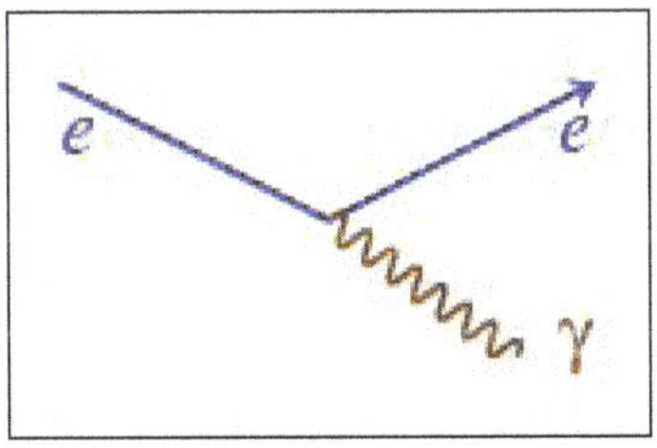

Feynman diagram for electron emitting a photon.

However, what happens to this equation when performing a gauge transformation? The equation changes, so that the solution for the field ψ should change as well. Indeed, ψ changes in the following way:

$$\Psi' = e^{-iq\Lambda}\psi, \quad \frac{\partial\psi'}{\partial t} = e^{-iq\Lambda}\left(\frac{\partial\psi}{\partial t} - iq\psi\frac{\partial\Lambda}{\partial t}\right)$$

Thus, the field ψ makes a rotation in the complex plane. This is closely related to a 'scale transformation', which would result if one were to remove the 'i' from $\Psi' = e^{-iq\Lambda}\psi$, $\frac{\partial\psi'}{\partial t} = e^{-iq\Lambda}\left(\frac{\partial\psi}{\partial t} - iq\psi\frac{\partial\Lambda}{\partial t}\right)$. It was Hermann Weyl who noted that this symmetry transformation simply redefines the scale of the field ψ, and introduced the word 'gauge' to describe this feature. The combinations $\vec{D}\psi = \left(\vec{\nabla} - iq\vec{A}\right)\psi$, $D_t\psi = \left(\frac{\partial}{\partial t} + iq\Phi\right)\psi$, are called covariant derivatives, because they are chosen in such a way that the derivatives of the function $\Lambda(\vec{x},t)$ cancel out in a gauge transformation:

$$\left(\vec{D}\psi\right)' = e^{-iq\Lambda}\left(\vec{\nabla} - iq\left(\vec{\nabla}\Lambda\right) - iq\left(\vec{A} - \vec{\nabla}\Lambda\right)\right)\psi = e^{-iq\Lambda}\left(\vec{D}\Psi\right),$$

$$(D_t\psi)' = e^{-iq\Lambda}\left(\frac{\partial}{\partial t} - iq\frac{\partial\Lambda}{\partial t} + iq\left(\Phi + \frac{\partial\Lambda}{\partial t}\right)\right)\psi = e^{-iq\Lambda}(D_t\Psi),$$

and this makes it easy to see that correctly $\Psi' = e^{-iq\Lambda}\psi$, $\frac{\partial\psi'}{\partial t} = e^{-iq\Lambda}\left(\frac{\partial\psi}{\partial t} - iq\psi\frac{\partial\Lambda}{\partial t}\right)$ describes the way ψ transforms under a local gauge transformation, obeying the same field equation both before and after the transformation (all terms in the equation are multiplied by the same exponential $e^{-iq\Lambda}$ so that that factor is immaterial).

The absolute value, $|\psi(\vec{x},t)|^2$ | does not change at all under a gauge transformation, and indeed this is the quantity that corresponds to something that is physically observable: it is the probability that a particle can be found at $(\vec{x},t)$. A rule of thumb is that local gauge invariance requires all derivatives in our equations to be replaced by covariant derivatives.

Yang-Mills Theory

In the 1950s, it was known that the field equations for the field of a proton, $P(\vec{x},t)$, and the field of a neutron $N(\vec{x},t)$, are such that one can rotate these fields in a complex two-dimensional space.

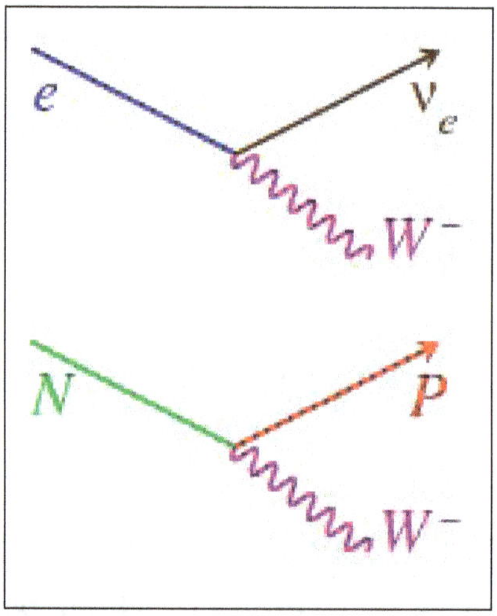

Feynman diagrams for emission of Yang-Mills photons.
Electron turning into en electron-neutrino neutron turning into proton.

$$\begin{pmatrix} P'(\vec{x},t) \\ N'(\vec{x},t) \end{pmatrix} = \begin{pmatrix} a & b \\ c & d \end{pmatrix}\begin{pmatrix} P(\vec{x},t) \\ N(\vec{x},t) \end{pmatrix},$$

Where, the matrix $U = \begin{pmatrix} a\,b \\ c\,d \end{pmatrix}$ may contain four arbitrary complex numbers, as long as it is unitary $(UU^{\dagger} = I)$, and usually, the determinant of U is restricted to be 1. Since these equations resemble the rotations one can perform in ordinary space, to describe spin of a particle, the symmetry in question here was called isospin.

In 1954, C.N. Yang and R.L. Mills published a very important idea. Could one modify the equations in such a way that these isospin rotations could be regarded as local gauge rotations? This would mean that, unlike the case that was known, the matrices U should be allowed to depend on space and time, just like the gauge generator $\Lambda(\vec{x},t)$ in electromagnetism. Yang and Mills were also inspired by the observation that Einstein's theory of gravity, General Relativity, also allows for transformations very similar to local gauge transformations: the replacement of the coordinate frame by other coordinates in an arbitrary, space-time dependent way.

To write down field equations for protons and neutrons, one needs the derivatives of these fields. The way these derivatives transform under a local gauge transformation implies that there will be terms containing the gradients $\vec{\nabla}U$ of the matrices U. To make the theory gauge-invariant, these gradients would have to be cancelled out, and in order to do that, Yang and Mills replaced the derivatives $\vec{\nabla}$ by covariant derivatives $\vec{D}=\vec{\nabla}-ig\vec{A}(\vec{x},t)$, as was done in electromagnetism. Here, however, the fields $\vec{A}$ had to be matrix-valued, just as the isospin U matrices:

$$\vec{A}=\begin{pmatrix} \vec{a}_{11} & \vec{a}_{12} \\ \vec{a}_{21} & \vec{a}_{22} \end{pmatrix}$$

$$\mathrm{Tr}\vec{A}=\vec{a}_{11}+\vec{a}_{22}=0, \quad \vec{a}_{11}=\vec{a}_{11}^{*} \quad \vec{a}_{21}=\vec{a}_{12}^{*}$$

Since the U matrices contain four coefficients with one constraint (the determinant has to be 1), one ends up with a set of three new vector fields (there are 3 independent real vectors in the matrix. At first sight, they appear to be the fields of a vector particle with isospin one. In practice, this should correspond to particles with one unit of spin (i.e., the particle rotates about its axis), and its electric charge could be neutral or one or minus one unit. Yang-Mills theory therefore predicts and describes a new type of particles with spin one that transmit a force not unlike the electro-magnetic force.

The fields that are equivalent to Maxwell's electric and magnetic fields are obtained by considering the commutator of two covariant derivatives:

$$\left[D_{\mu},D_{\nu}\right]=D_{\mu}D_{\nu}-D_{\nu}D_{\mu}=-ig\left(\partial_{\mu}A_{\nu}-\partial_{\nu}A_{\mu}-ig\left[A_{\mu},A_{\nu}\right]\right)=-igF_{\mu\nu},$$

Where the indices take the values $\mu, \nu=0,1,2,3$, with 0 referring to the time-component.

Since $F_{\mu\nu}=-F_{\nu\mu}$, this tensor has 6 independent components, three forming an electric vector field, and three a magnetic field. Each of these components is also a matrix. The commutator, $[A_{\mu},A_{\nu}]$ is a new, non-linear term, which makes the Yang-Mills equations a lot more complicated than the Maxwell system.

In other respects, the Yang-Mills particles, being the energy quanta of the Yang-Mills fields, are similar to photons, the quanta of light. Yang-Mills particles also carry no intrinsic mass, and travel with the speed of light. Indeed, these features were at first reasons to dismiss this theory, because massless particles of this sort should have been detected long ago, whereas they were conspicuously absent.

The Brout-Englert-Higgs Mechanism

The theory was revived when it was combined with spontaneous breakdown of local gauge symmetry, also known as the Brout-Englert-Higgs mechanism. Consider a scalar (spinless) particle described by a field $\phi(\vec{x},t)$. This field is assumed to be a vector field, in the sense that it undergoes some rotation when a gauge transformation is performed. In practice this means that the particle carries one or several kinds of charges that make it sensitive to the Yang-Mills force, and often it has several components, which means there are various species of this particle. Such particles must obey Bose-Einstein statistics, which implies that it can undergo Bose-Einstein condensation. In terms of its field φ this means the following:

In the vacuum the field φ takes a non-vanishing value F. This is usually written as:

$$\langle \phi(\vec{x},t) \rangle = F$$

After a local gauge transformation, this would look like:

$$\langle \phi'(\vec{x},t) \rangle = U(\vec{x},t)F$$

where $U(\vec{x},t)$ is a matrix field representing the local gauge transformation.

It is often said that, therefore, the vacuum is not gauge-invariant, but, strictly speaking, this is not correct. The situation described by $\langle \phi'(\vec{x},t) \rangle = U(\vec{x},t)F$ is the same vacuum as $\langle \phi(\vec{x},t) \rangle = F$; it is only described differently. However, this property of the vacuum does have important consequences. Due to the fact that the rotated field now describes the same situation as the previous value, there is no different physical particle associated to the rotated field. Only the length of the vector φ has physical significance. This length is gauge-invariant. Therefore, only the length of the vector φ is associated to one type of particle, which must be neutral for the Yang-Mills forces. This particle is now called the Higgs particle.

As the Higgs field is a constant source for the Yang-Mills field strength, the Yang-Mills field equations are modified by it. Due to the Higgs field, the Yang-Mills "photons" described by the Yang-Mills field $A_\mu(\vec{x},t)$ get a mass. Massless photons can only have two helicity states, that is, they can spin only in two directions. This is related to the fact that light can be polarized in exactly two directions. Massive photons (particles with non-vanishing mass and with one unit of spin), can always spin in three directions. This third rotation mode is now provided by the Higgs field, which itself loses several of its physical components. The total number of physical field components stays the same before and after the Brout-Englert-Higgs mechanism. A further consequence of this effect on the Yang-Mills field is that the force transmitted by the massive photons is a short-range one (the range of the force being inversely proportional to the mass of the photon).

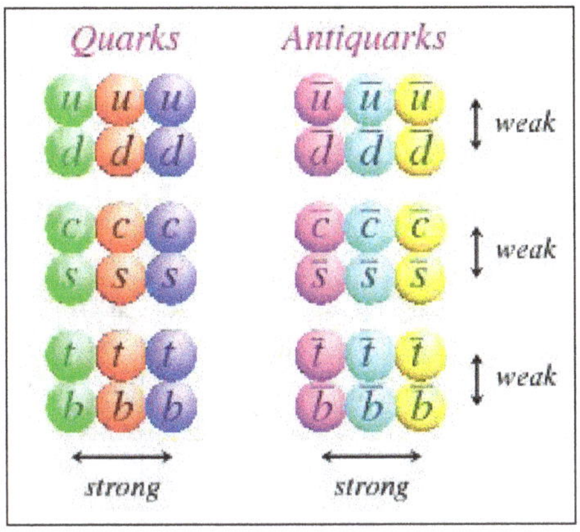

The six flavors and three colors of quarks and their antiparticles. Arrows show the weak and the strong transitions.

The weak interactions could now be successfully described by a Yang-Mills theory. The set of local gauge transformations forms the mathematical group $SU(2)\times U(1)$. This group generates 4 species of photons (3 for SU(2) and 1 for U(1)). The Brout-Englert-Higgs mechanism breaks this group down in such a way that a subgroup of the form U(1) remains. This is the electromagnetic theory, with just one photon. The other three photons become massive; they are responsible for the weak interactions, which in practice appear to be weak just because these forces have a very short range. With respect to electromagnetism, two of these intermediate vector bosons, $W^{\pm}$, are electrically charged, and a third, Z^{o} , is electrically neutral. When the latter's existence was derived from group theoretical arguments, this gave rise to the prediction of a hitherto unnoticed form of the weak interaction: the neutral current interaction. This theory, that combines electromagnetism and the weak force into one, is called the electro-weak theory, and it was the first fully renormalizable theory for the weak force.

Quantum Chromodynamics

When it was understood that the weak interactions, together with the electromagnetic ones, can be ascribed to a Yang-Mills gauge theory, the question was asked how to address the strong force, a very strong force with relatively short range of action, which controls the behavior of the hadronic particles such as the nucleons and the pions. These particles behave as if built from subunits, called quarks. Three varieties of quarks were known (up, down, and strange), and three more would be discovered later (charm, top, and bottom). These quarks have the peculiar property that they permanently stick together either in triplets, or one quark sticks together with one anti-quark. Yet when they approach one another very closely, they begin to behave more freely as individuals.

These features we now understand as, again, being due to a Yang-Mills gauge theory. Here, we have the mathematical group SU(3) as local gauge group, while now the symmetry is not affected by any Brout-Englert-Higgs mechanism. Due to the non-linear nature of the Yang-Mills field, it self-interacts, which forces the fields to come in patterns quite different from the electromagnetic case: vortex lines are formed, which form unbreakable bonds between quarks. At close distances, the Yang-Mills force becomes weak, and this is a feature that can be derived in an elementary way using perturbation expansions, but it is a property of the quantized Yang-Mills system that hitherto had been thought to be impossible for any quantum field theory, called asymptotic freedom.

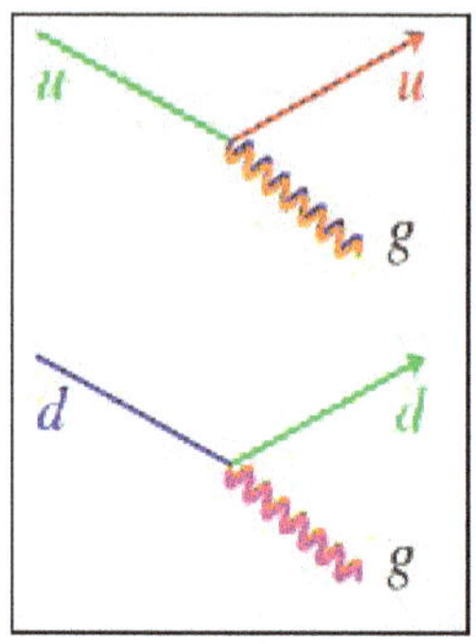

Feynman diagrams for emission of QCD gluons. Quarks change color, but their flavor stays the same: u stays u and d stays d.

SU(3) implies that every species of quark comes in three types, referred to as color: they are "red", "green" or "blue". The field of a quark is therefore a 3-component vector in an internal 'color' space. Yang-Mills gauge transformations rotate this vector in color space. The Yang-Mills fields themselves form 3 by 3 matrices, with one constraint (since the determinant of the Yang-Mills gauge rotation matrices must be kept equal to one). Therefore, the Yang-Mills field has 8 colored photon-like particles, called gluons. Anti-quarks carry the conjugate colors ("cyan", "magenta" or "yellow"). The theory is now called Quantum chromodynamics (QCD). It is also a renormalizable theory.

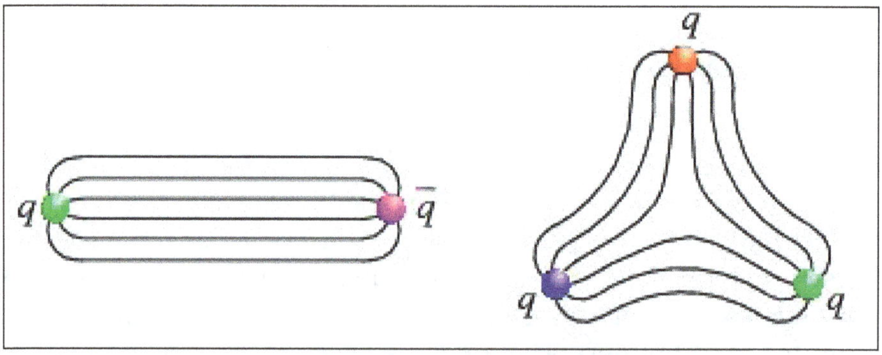

The quantum chromodynamical fields form vortices that keep quarks and antiquarks (left) or three-quark systems (right) permanently confined.

The gluons effectively keep the quarks together in such a way that their colors add up to a total that is color-neutral ("white" or a "shade of gray"). This is why either three quarks or one quark and one anti-quark can sit together to form a physically observable particle (a hadron). This property of the theory is called permanent quark confinement. Because of the strongly non-linear nature of the fields, quark confinement is in fact quite difficult to prove, whereas the property of asymptotic freedom can be demonstrated exactly.

The Lagrangian

One cannot choose all field equations at will. They must obey conditions such as energy conservation. This implies that there is an action principle (action = reaction), and this principle is most conveniently expressed by writing the Lagrangian for the theory. The Lagrangian (more precisely, Lagrange density) $L(\vec{x},t)$ is an expression in terms of the fields of the system. For a real scalar field Φ it is:

$$L=-\frac{1}{2}\left(\left(\vec{D}\Phi\right)^2-\left(D_t\Phi\right)^2+m^2\Phi^2\right)$$

And for the Maxwell fields it is:

$$L=-\frac{1}{2}\left(\vec{E}^2-\vec{B}^2\right)=-\frac{1}{4}\sum_{\mu,\nu}F_{\mu\nu}F_{\mu\nu},$$

Where the summation is the Lorentz covariant summation over the Lorentz indices μ, ν. The field equations can all be derived from this expression by demanding that the action integral,

$$S=\int d^3\vec{x}\,dt\,L(\vec{x},t)$$

Where L is the sum of the Lagrangians of all fields in the system, be stationary under all infinitesimal variations of these fields. This is called the Euler-Lagrange principle, and the equations are the Euler-Lagrange equations. For gauge theories this generalizes directly, one writes,

$$L=-\frac{1}{4}\mathrm{Tr}\sum_{\mu,\nu}F_{\mu\nu}F_{\mu\nu},$$

Using the expression $\left[D_\mu,D_\nu\right]=D_\mu D_\nu-D_\nu D_\mu=-ig\left(\partial_\mu A_\nu-\partial_\nu A_\mu-ig\left[A_\mu,A_\nu\right]\right)=-igF_{\mu\nu},$ for the gauge fields $F_{\mu\nu}$, and adds all terms associated to the other fields that are introduced. All symmetries of the theory are the symmetries of the Lagrangian, and the dimensionality of all coupling strengths can easily be read off from the Lagrangian as well, which is of importance for the renormalization procedure.

Renormalization and Anomalies

According to the laws of quantum mechanics, the energy in a field consists of energy packets, and these energy packets are in fact the particles associated to the field. Quantum mechanics gives extremely precise prescriptions on how these particles interact, as soon as the field equations are known and can be given in the form of a Lagrangian. The theory is then called quantum field theory (QFT), and it explains not only how forces are transmitted by the exchange of particles, but it also states that multiple exchanges should occur. In many older theories, these multiple exchange gave rise to difficulties: their effects seem to be unbounded, or infinite. In a gauge theory, however, the small distance structure is very precisely prescribed by the requirement of gauge-invariance. In such a theory one can combine the infinite effects of the multiple exchanges with redefinitions of masses and charges of the particles involved. This procedure is called renormalization. In 3 space and 1 time dimension, most gauge theories are renormalizable. This allows us to compute the effects of multiple particle exchanges to high accuracy, thus allowing for detailed comparison with experimental data.

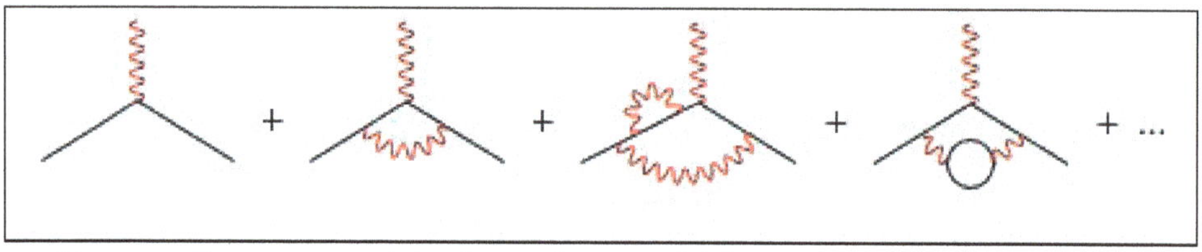

Feynman diagrams containing loops, due to multiple particle exchanges. The loops often generate infinite expressions.

Renormalization requires that masses and coupling strengths of particles be defined very carefully. If all coupling parameters of a theory are given a mass-dimensionality that is zero or positive, the number of divergent expressions stays under control. Usually, requiring the theory to remain gauge invariant throughout the renormalization procedure leaves no ambiguity for the definitions. However, unambiguous, gauge invariant definitions exist at all, since gauge invariance has to hold for all interactions, whereas only a few infinite expressions can be replaced by finite ones.

The proof that showed how and why unambiguous renormalized expressions can be obtained, could be most elegantly obtained by realizing that gauge theories can be formulated in any number of space-time dimensions. It was even possible to define all Feynman diagrams unambiguously for theories in spaces where the dimensions are $3-\epsilon$, where ϵ is an infinitesimal quantity. Taking the limit $\epsilon \to 0$ requires the subtraction of poles of the form C_n / ϵ^n from the original, "bare" mass and coupling parameters. The result is a set of unique, finite and gauge invariant expressions. In practice, it was found that this procedure, called dimensional regularization and renormalization is also convenient for carrying out technically complicated calculations of loop diagrams.

However, there is a special case where extension to dimensions different from the canonical one is impossible. This is when fermionic particles exhibit chiral symmetry. Chiral symmetry is a symmetry that distinguishes left-rotating from right rotating particles, and indeed it plays a crucial role in the Standard Model. Chiral symmetry is only possible if space is 3 dimensional, and so does not allow for dimensional renormalization. Indeed, sometimes chiral symmetry cannot be preserved when renormalizing the theory. An anomaly occurs, called chiral anomaly. It was first discovered when a calculation of the $\pi_0 \rightarrow \gamma\gamma$ decay amplitude gave answers that did not follow the expected symmetry pattern.

Since the gauge symmetries of the Standard Model do distinguish left rotating from right rotating particles (in particular, only left-rotating neutrinos are produced in a weak interaction), anomalies were a big concern. It so happens, however, that all anomalous amplitudes that would jeopardize gauge invariance and hence the self-consistency of our equations, all cancel out. This is related to the fact that certain "grand unified" extensions of the Standard Model are based on anomaly free gauge groups.

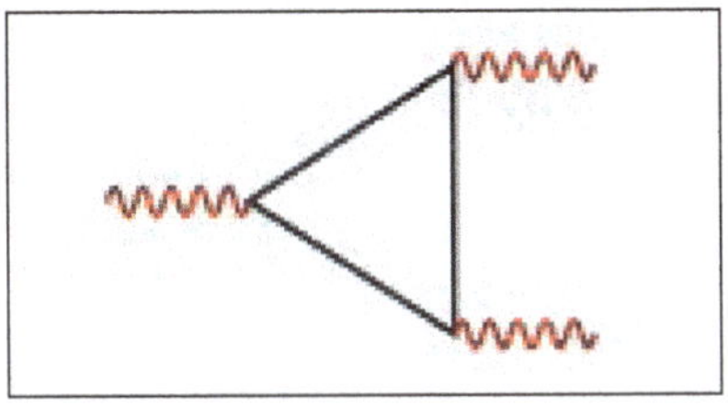

The diagram where a fermionic particle forms a closed triangle, coupling to three gauge particles, is the main source of anomalies.

The anomaly has a direct physical implication. A topologically twisted field configuration called the instanton (because it represents an event at a given instant in time), represents exactly the gauge field configuration where the anomaly is maximal. It causes a violation of the conservation of some of the gauge charges. When there is an anomaly, at least one of the charges involved cannot be a gauge charge, but must be a charge to which no gauge field is coupled, like baryonic charge. Indeed, in the electroweak theory, instantons trigger the violation of the conservation laws of baryons. It is now believed that this might explain the imbalance between matter and antimatter that must have arisen during early phases of the Universe.

Higgs Boson Theory

Higgs boson is also called Higgs particle, particle that is the carrier particle, or boson, of the Higgs field, a field that permeates space and endows all elementary subatomic particles with mass through its interactions with them.

The Higgs field is different from other fundamental fields such as the electromagnetic field—that underlie the basic forces between particles. First, it is a scalar field; i.e., it has magnitude but no direction. This implies that its carrier, the Higgs boson, has an intrinsic angular momentum, or spin, of 0, unlike the carriers of the force fields, which have spin. Second, the Higgs field has the unusual property that its energy is higher when the field is zero than when it is nonzero. The elementary particles therefore acquired their masses through interactions with a nonzero Higgs field only when the universe cooled and became less energetic in the aftermath of the big bang (the hypothetical primal explosion in which the universe originated). The variety of masses characterizing the elementary subatomic particles arises because different particles have different strengths of interaction with the Higgs field.

Quantum Field Theory

Quantization of the Free Scalar Field

A free scalar field can be described by the Lagrangian:

$$L = \int d^3x \mathscr{L}$$

Where,

$$\mathscr{L} = \frac{1}{2}\partial_\mu \varphi \partial^\mu \varphi - \frac{1}{2}m^2\varphi^2$$

$$= \frac{1}{2}\phi^2 - \frac{1}{2}\nabla_{i\phi}\nabla_{i\phi} - \frac{1}{2}m^2\phi^2$$

Our goal is to "quantize" this theory, in the sense of developing a quantum theory that corresponds to the classical theory described by the above Lagrangian.

1. Canonical Quantization: Here we will use the method of canonical quantization, which we assume is already familiar to you in the context of quantum mechanics. Specifically, suppose that we were given a Lagrangian with a discrete number of dynamical variables q_i:

$$L = L(q_i, \dot{q}_i, t).$$

The canonical momenta would then be defined by:

$$p_i \equiv \frac{\partial L}{\partial \dot{q}i},$$

And the Hamiltonian would be given by,

$$H = \sum_i p_i \dot{q}_i - L$$

A quantum theory corresponding to this classical theory could then be constructed by promoting each q_i and p_i to an operator on a Hilbert space, and insisting on the canonical commutation relations:

$$\left[q_i, p_j\right] = i\hbar\delta_{ij} .$$

For most of this course we will use units for which $\hbar \equiv 1$, but for now we will leave the $\hbar$ in the equations. The Hamiltonian $H(p_i, q_i)$ is then also an operator on the Hilbert space, and in the Schrodinger picture the physical states evolve according to the Schrodinger equation:

$$i\hbar\frac{\partial}{\partial t}|\psi(t)\rangle = H|\psi(t)\rangle$$

If H is independent of time, $i\hbar\frac{\partial}{\partial t}|\psi(t)\rangle = H|\psi(t)\rangle$ has the formal solution:

$$|\psi(t)\rangle = e^{-iHt/\hbar}|\psi(o)\rangle$$

Given any operator $\mathcal{O}$, its expectation value in the state $|\psi(t)\rangle$ is then given by:

$$\langle\psi(t)|\mathcal{O}|\psi(t)\rangle = \langle\psi(o)|e^{iHt/\hbar}\mathcal{O}e^{-iHt/\hbar}|\psi(o)\rangle$$

This equation leads naturally to the Heisenberg picture description, in which the states are treated as time-independent, and all of the time dependence is incorporated into the evolution of the operators:

$$\mathcal{O}(t) = e^{iHt/\hbar}\ \mathcal{O}\ e^{-iHt/\hbar}$$

Field Quantization by Lattice Approximation

To quantize the classical field theory of $\mathcal{L} = \frac{1}{2}\partial_\mu\varphi\partial^\mu\varphi - \frac{1}{2}m^2\varphi^2$ we can begin by quantizing a lattice version of the theory.That is, we can replace the continuous space by a cubic lattice of closely spaced grid points, with a lattice spacing a, and we can truncate the space to a finite region. The system then reduces to one with a discrete number of dynamical variables, exactly like the systems that we already know how to quantize. Then if we can take the limit as the lattice spacing a approaches zero and the volume approaches infinity, the quantization of the field theory can be completed. We will see later that the $a \to o$ limit is problematic for interacting theories, but we will see here that this program can be carried out easily for the free theory.

When we replace the continuous space by a finite lattice of points, we can label each lattice site with an index k. In a fully detailed lattice description we would probably label each lattice site with a triplet of integers representing the x, y, and z coordinates of the site, but for present purposes it will suffice to imagine simply numbering all the lattice sites from 1 to N, where N is the total number of sites. The field $\phi(\vec{x},t)$ is then replaced by a set of dynamical variables $\phi_k(t)$ where one can think of $\phi_k(t)$ as representing the average value of $\phi(\vec{x},t)$ in a cube of size a surrounding the lattice site k. The Lagrangian of $L=\int d^3x\mathscr{L}$ and $\mathscr{L}=\frac{1}{2}\partial_\mu\varphi\partial^\mu\varphi-\frac{1}{2}m^2\varphi^2$ is then replaced by:

$$L=\sum_k \mathscr{L}_k \Delta V$$

Where,

$$\Delta V = a^3$$

and,

$$L_k = \frac{1}{2}\dot{\phi}_k^2 - \frac{1}{2}\nabla_{i\phi k}\nabla_{i\phi k} - \frac{1}{2}m^2\phi_k^2$$

Here the lattice derivative $\nabla_{i\,k}$ is defined by,

$$\nabla_{i\phi k} \equiv \frac{\phi_{k'(k,i)}-\phi_k}{a}$$

where k' (k, i) denotes the lattice site that is a distance a in the *i*th direction from the lattice site k. The canonical momenta are then given by,

$$\pi_k \equiv \frac{p_k}{\Delta V} = \frac{\partial \mathscr{L}_k}{\partial \dot{\phi}_k} = \dot{\phi}_k$$

Following $H=\sum_i p_i\dot{q}_i - L$, the Hamiltonian is then,

$$H=\sum_k p_k\dot{\phi}_k - L = \sum_k \left[\pi_k\dot{\phi}_k - \mathscr{L}_k\right]\Delta V$$

The canonical commutation relations become:

$$[\phi_{k'},\phi_k]=0,\ [P_{k'},P_k]=0 \text{ and } [\phi_{k'},p_k]=i\hbar\delta_{k'k}$$

In terms of the canonical momentum densities,

$$[\phi_{k'},\phi_k]=0\ ,[\pi_{k'},\pi_k]=0,\ \text{and}\ [\phi_{k'},\pi_k]=\frac{i\hbar\delta_{k'k}}{\Delta V}.$$

Although we have not yet constructed the full theory, it is not too early to write down the continuum limit of these defining equations. The continuum canonical momentum density becomes,

$$\pi(\vec{x},t)=\frac{\partial L}{\partial\dot{\phi}(\vec{x},t)}=\dot{\phi}(\vec{x},t),$$

and the Hamiltonian becomes,

$$H=\int d^3x\left[\pi\dot{\phi}-\mathscr{L}\right].$$

The trivial canonical commutation relations carry over trivially:

$$\left[\phi\left(\vec{x}',t\right),\phi(\vec{x},t)\right]=0 \text{ and } \left[\pi\left(\vec{x}',t\right),\pi(x,t)\right]=0$$

obviously. For the nontrivial commutation relation, the result will be clearest if we first rewrite the last equation in $[\phi_{k'},\phi_k]=0$,$[\pi_{k'},\pi_k]=0$, and $[\phi_{k'},\pi_k]=\frac{i\hbar\delta_{k'k}}{\Delta V}$. as a sum which will become an integral in the limit. If we let $\mathcal{R}$ denote a region of the lattice:

$$\sum_{k\in\mathcal{R}}[\phi_{k'},\pi_k]\Delta V=i\hbar\sum_{k\in\mathcal{R}}\delta_{k'k}=i\hbar\begin{cases}1 & \text{if } k'\in\mathcal{R}\\ 0 & \text{otherwise.}\end{cases}$$

In the continuum limit,

$$\sum_{k\in\mathcal{R}}[\phi_{k'},\pi_k]\Delta V$$

Clearly approaches,

$$\int_{x\in\mathcal{R}}d^3x\left[\phi\left(\vec{x}',t\right),\pi(\vec{x},t)\right]$$

so $\sum_{k\in\mathcal{R}}[\phi_{k'},\pi_k]\Delta V=i\hbar\sum_{k\in\mathcal{R}}\delta_{k'k}=i\hbar\begin{cases}1 & \text{if } k'\in\mathcal{R}\\ 0 & \text{otherwise.}\end{cases}$ becomes:

$$\int_{x\in\mathcal{R}}d^3x\left[\phi\left(\vec{x}',t\right),\pi(\vec{x},t)\right]=i\hbar\begin{cases}1 & \text{if } k'\in\mathcal{R}\\ 0 & \text{otherwise.}\end{cases}$$

This relationship can be expressed more conveniently by introducing the Dirac delta function $\delta^3(\vec{x})$, which is defined by its integral*:

$$\int_{x\in\mathcal{R}}d^3x f(\vec{x})\delta\left(\vec{x}-\vec{x}'\right)=\begin{cases}f(\vec{x}') & \text{if } x'\in\mathcal{R}\\ 0 & \text{otherwise.}\end{cases}$$

Given this definition, $\int_{x\in\mathcal{R}}d^3x\left[\phi\left(\vec{x}',t\right),\pi(\vec{x},t)\right]=i\hbar\begin{cases}1 & \text{if } k'\in\mathcal{R}\\ 0 & \text{otherwise}\end{cases}$ can be rewritten as:

$$\left[\phi(\vec{x}',t),\pi(\vec{x},t)\right]=i\hbar\delta(\vec{x}-\vec{x}').$$

The delta function is symmetric, so $\delta(\vec{x}-\vec{x}')=\delta(\vec{x}'-\vec{x})$.

Simple Harmonic Oscillator

We will soon see that each Fourier component of a scalar field obeys the equations of a harmonic oscillator, so it is useful to review the quantum mechanics of a harmonic oscillator before we proceed. Consider the Lagrangian for a simple harmonic oscillator, which can be written as:

$$L = \frac{1}{2}\dot{q}^2 - \frac{1}{2}\omega^2 q^2$$

Where we have adopted units for which the mass m of the harmonic oscillator is one. The canonical momentum is,

$$p = \frac{\partial L}{\partial \dot{q}} = \dot{q}$$

and the Hamiltonian is,

$$H = p\dot{q} - L = \frac{1}{2}p^2 + \frac{1}{2}\omega^2 q^2$$

The canonical commutation relation is,

$$[q, p] = i\hbar$$

Now we can define creation and annihilation operators,

$$a = \sqrt{\frac{\omega}{2\hbar}}q + \frac{i}{2\hbar\omega}p,\ a^{\dagger} = \sqrt{\frac{\omega}{2\hbar}}q - \frac{i}{2\hbar\omega}p,$$

So that, $\left[a, a^{\dagger}\right] = 1.$

The Hamiltonian can be rewritten as,

$$H = \hbar\omega\left(a^{\dagger}a + \frac{1}{2}\right)$$

The commutators of H with the creation and annihilation operators are,

$$\left[H, a^{\dagger}\right] = \hbar\omega a^{\dagger},\ (H, a) = -\hbar\omega a,$$

from which it follows that the result of applying $a^{\dagger}$ to an eigenstate of H is to produce an eigenstate of H with an eigenvalue higher than the original by $\hbar\omega$, while a acts on an eigenstate of H to lower the eigenvalue by $\hbar\omega$. The ground state of H therefore satisfies,

$$a|o\rangle = o$$

While the (normalized) nth excited state can be written as,

$$|n\rangle = \frac{1}{\sqrt{n!}}\left(a^{\dagger}\right)^{n}|n\rangle$$

where the eigenvalue (energy) is,

$$E_n = \left(n + \frac{1}{2}\right)\hbar\omega$$

Eqs. $a = \sqrt{\frac{\omega}{2\hbar}}q + \frac{i}{2\hbar\omega}p,\ a^{\dagger} = \sqrt{\frac{\omega}{2\hbar}}q - \frac{i}{2\hbar\omega}p,$ can be solved for q and p, giving,

$$q = \sqrt{\frac{\hbar}{2\omega}}\left(a + a^{\dagger}\right), p = -i\sqrt{\frac{\hbar\omega}{2}}\left(a - a^{\dagger}\right).$$

In the Heisenberg picture,

$$q(t) = e^{iHt/\hbar}\, q\, e^{-iHt/\hbar}$$

$$= \sqrt{\frac{\hbar}{2\omega}} e^{iHt/\hbar}\left(a + a^{\dagger}\right)e^{-iHt/\hbar}$$

$$= \sqrt{\frac{\hbar}{2\omega}}\left(ae^{-i\omega t} + a^{\dagger}e^{i\omega t}\right)$$

and,

$$p(t) = -\sqrt{\frac{\hbar\omega}{2}}\left(ae^{-i\omega t} - a^{\dagger}e^{i\omega t}\right).$$

Quantization of the Scalar Field

We have used the continuum limit of the lattice version of the theory to obtain the key results,

$$\pi(\vec{x},t) = \frac{\partial L}{\partial\dot{\phi}(\vec{x},t)} = \dot{\phi}(\vec{x},t)$$

$$H = \int d^3x\left[\pi\dot{\phi} - \mathscr{L}\right]$$

$$\left[\phi\left(\vec{x}',t\right), \phi(\vec{x},t)\right] = 0 \text{ and } \left[\pi\left(\vec{x}',t\right), \pi(x,t)\right] = 0$$

$$\left[\phi(\vec{x}',t), \pi(\vec{x},t)\right] = i\hbar\delta(\vec{x} - \vec{x}')$$

Having done this, we can now proceed with the continuum theory directly. We view $\phi(\vec{x},t)$ as a collection of dynamical variables, in the classical theory, which have been promoted to operators in the quantum theory, exactly as we did for the discrete system. We can then use the Fourier transform to define convenient linear combinations of these operators:

$$\tilde{\phi}(\vec{k},t) \equiv \int d^3x e^{-i\vec{k}\cdot\vec{x}} \phi(\vec{x},t)$$

and

$$\tilde{\pi}(\vec{k},t) \equiv \int d^3x e^{-i\vec{k}\cdot\vec{x}}\ \pi(\vec{x},t),$$

So that the Fourier inversion theorem implies that,

$$\phi(\vec{x},t) = \int \frac{d^3k}{(2\pi)^3} e^{i\vec{k}\cdot\vec{x}}\ \tilde{\phi}(\vec{k},t)$$

$$\pi(\vec{x},t) = \int \frac{d^3k}{(2\pi)^3} e^{i\vec{k}\cdot\vec{x}}\ \tilde{\pi}(\vec{k},t).$$

The fact that $\phi(\vec{x},t)$ is a real classical variable and hence a Hermitian quantum operator implies that,

$$\tilde{\phi}(-\vec{k},t) = \phi^{\dagger}(\vec{k},t)\ ,$$

With a similar relation for $\pi(\vec{k},t)$.

The Heisenberg equations of motion for $\phi(\vec{x},t)$ are the same as the classical equations of motion:

$$\frac{\partial^2\phi}{\partial t^2} - \vec{\nabla}^2\phi + m^2\phi = 0$$

Since we have not set $\hbar$ to one, we should keep in mind that m in this equation has the units of an inverse length, and not a mass. $\frac{\partial^2\phi}{\partial t^2} - \vec{\nabla}^2\phi + m^2\phi = 0$ implies that the Fourier transform field obeys,

$$\frac{\partial^2\tilde{\phi}}{\partial t^2}(\vec{k},t) + \left(\vec{k}^2 + m^2\right)\tilde{\phi}(\vec{k},t) = 0$$

The general solution to this equation can be written as:

$$\tilde{\phi}(\tilde{k},t) = \phi 1(\vec{k}) e^{-i\omega_p t} + \phi_2(\vec{k}\ ) e^{i\omega_p t}$$

$$\omega_p = \sqrt{\vec{k}^2 + m^2}$$

The reality condition $\tilde{\phi}(-\vec{k},t) = \phi^{\dagger}(\vec{k},t)$, implies that,

$$\phi_2(\vec{k}) = \phi_1^{\dagger}(-\vec{k}),$$

So,

$$\tilde{\phi}(\vec{k},t) = \phi 1(\vec{k})e^{-i\omega_p t} + \phi_1^{\dagger}(-\vec{k})e^{i\omega_p t}$$

The relation $\pi = \dot{\phi}$ then implies that,

$$\tilde{\pi}(\vec{k},t) = -i\omega_p \phi 1(\vec{k})e^{-i\omega_p t} + i\omega_p \phi_1^{\dagger}(-\vec{k})e^{i\omega_p t}$$

And then $\tilde{\phi}(\vec{k},t) = \phi 1(\vec{k})e^{-i\omega_p t} + \phi_1^{\dagger}(-\vec{k})e^{i\omega_p t}$ and $\tilde{\pi}(\vec{k},t) = -i\omega_p \phi 1(\vec{k})e^{-i\omega_p t} + i\omega_p \phi_1^{\dagger}(-\vec{k})e^{i\omega_p t}$ can be solved simultaneously to give an expression for $\phi_1(\vec{k})$,

$$\phi_1(\vec{k}) = \frac{1}{2}\left[\tilde{\phi}(\vec{k},t) + \frac{i}{\omega_p}\tilde{\pi}(\vec{k},t)\right]e^{i\omega_p t}$$

$$= \frac{1}{2}\int d^3x e^{-i(\vec{k}\cdot\vec{x}-\omega_p t)}\left[\phi(\vec{x},t) + \frac{i}{\omega_p}\pi(\vec{x},t)\right]$$

Where, the second line was obtained by using $\tilde{\phi}(\vec{k},t) \equiv \int d^3x e^{-i\vec{k}\cdot\vec{x}}\phi(\vec{x},t)$ and $\tilde{\pi}(\vec{k},t) \equiv \int d^3x e^{-i\vec{k}\cdot\vec{x}}\,\pi(\vec{x},t)$, although the right-hand side contains quantities that depend explicitly on t, $\tilde{\phi}(\vec{k},t) = \phi 1(\vec{k})e^{-i\omega_p t} + \phi_1^{\dagger}(-\vec{k})e^{i\omega_p t}$ And $\tilde{\pi}(\vec{k},t) = -i\omega_p \phi 1(\vec{k})e^{-i\omega_p t} + i\omega_p \phi_1^{\dagger}(-\vec{k})e^{i\omega_p t}$ guarantee that the full expression is independent of time. By comparing with , $a = \sqrt{\frac{\omega}{2\hbar}}q + \frac{i}{2\hbar\omega}p$, $a^{\dagger} = \sqrt{\frac{\omega}{2\hbar}}q - \frac{i}{2\hbar\omega}p$, one sees that $\phi_1(\vec{k})$ bears some resemblance to an annihilation operator. To test this hypothesis, we can compute, $\left[\phi_1(\vec{k}),\phi_1(\vec{q})\right]$ and $\left[\phi_1(\vec{k}),\phi_1^{\dagger}(\vec{q})\right]$.

Using $\phi_1(\vec{k}) = \frac{1}{2}\left[\tilde{\phi}(\vec{k},t) + \frac{i}{\omega_p}\tilde{\pi}(\vec{k},t)\right]e^{i\omega_p t}$, $= \frac{1}{2}\int d^3x e^{-i(\vec{k}\cdot\vec{x}-\omega_p t)}\left[\phi(\vec{x},t) + \frac{i}{\omega_p}\pi(\vec{x},t)\right]$ and the canonical commutation relations,

$$\left[\phi_1\left(\vec{k}\right),\phi_1\left(\vec{q}\right)\right]=\frac{1}{4}\int d^3x\int d^3y e^{-i\left(\vec{k}\cdot\vec{x}-\omega_k t\right)}e^{-1\left(\vec{q}\cdot\vec{y}-\omega_q t\right)}$$

$$\times\left[\phi\left(\vec{x},t\right)+\frac{i}{\omega_k}\pi\left(\vec{x},t\right),\phi\left(\vec{y},t\right)+\frac{i}{\omega_q}\pi\left(\vec{y},t\right)\right]$$

$$=\frac{1}{4}\int d^3x\int d^3y e^{-i\left(\vec{k}\cdot\vec{x}-\omega_k t\right)}e^{-1\left(\vec{q}\cdot\vec{y}-\omega_q t\right)}$$

$$\times\left\{\frac{i}{\omega q}i\hbar\delta^3\left(\vec{x}-\vec{y}\right)-\frac{i}{\omega_k}i\hbar\delta^3\left(\vec{x}-\vec{y}\right)\right\}$$

$$=\frac{\hbar}{4}\int d^3x e^{-i\left(\left(\vec{k}+\vec{q}\right).\vec{x}-\left(\omega_k+\omega_q\right)t\right)}\left\{\frac{1}{\omega_k}-\frac{1}{\omega_q}\right\}$$

We now use the identity,

$$\int d^3x e^{-i\left(\vec{k}+\vec{q}\right)\cdot\vec{x}}=\left(2\pi\right)^3\delta^3\left(\vec{k}+\vec{q}\right)$$

Which implies that the only possible nonzero contribution arises when $\vec{q}=-\vec{k}$, but in that case the factor in curly brackets vanishes. Thus,

$$\left\{\phi_1\left(\vec{k}\right),\phi_1\left(\vec{q}\right)\right\}=0$$

As we would expect if $\phi_1\left(\vec{k}\right)$ is proportional to an annihilation operator. Using the same techniques, we can now calculate:

$$\left[\phi_1\left(\vec{k}\right),\phi_1\left(\vec{q}\right)\right]=\frac{1}{4}\int d^3x\int d^3y e^{-i\left(\vec{k}\cdot\vec{x}-\omega_k t\right)}e^{-1\left(\vec{q}\cdot\vec{y}-\omega_q t\right)}$$

$$\times\left[\phi\left(\vec{x},t\right)+\frac{i}{\omega_k}\pi\left(\vec{x},t\right),\phi\left(\vec{y},t\right)+\frac{i}{\omega_q}\pi\left(\vec{y},t\right)\right]$$

$$=\frac{1}{4}\int d^3x\int d^3y e^{-i\left(\vec{k}\cdot\vec{x}-\omega_k t\right)}e^{-1\left(\vec{q}\cdot\vec{y}-\omega_q t\right)}$$

$$\times\left\{\frac{i}{\omega q}i\hbar\delta^3\left(\vec{x}-\vec{y}\right)-\frac{i}{\omega_k}i\hbar\delta^3\left(\vec{x}-\vec{y}\right)\right\}$$

$$=\frac{\hbar}{4}\int d^3x e^{-i\left(\left(\vec{k}+\vec{q}\right).\vec{x}-\left(\omega_k+\omega_q\right)t\right)}\left\{\frac{1}{\omega_k}-\frac{1}{\omega_q}\right\}$$

$$=\frac{\hbar}{4}(2\pi)^3\delta^3(\vec{k}-\vec{q})\frac{2}{\omega_k}=\frac{\hbar}{2\omega_k}(2\pi)^3\delta^3(\vec{k}-\vec{q})$$

The standard (continuum) definition of the annihilation operator is then given by,

$$a(\vec{k})=\sqrt{\frac{2\omega_k}{\hbar}}\phi_1(\vec{k})$$

So the commutation relations become,

$$\left[a(\vec{k}),a(\vec{q})\right]=0,\left[a(\vec{k}),a^\dagger(\vec{q})\right]=(2\pi)^3\delta^3(\vec{k}-\vec{q})$$

The field operator can now be expressed in terms of creation and annihilation operators:

$$\phi(\vec{x},t)=\int\frac{d^3k}{(2\pi)^3}\sqrt{\frac{\hbar}{2\omega_k}}\left\{a(\vec{k})e^{i(\vec{k}\cdot x-\omega_k t)}+a^\dagger(\vec{k})e^{-i(\vec{k}\cdot\vec{x}-\omega_k t)}\right\}$$

In the second term we have changed variables of integration $\vec{k}\to-\vec{k}$, to write in the form that is shown. The canonical momentum density is then given by,

$$\pi(\vec{x},t)=\dot{\phi}(\vec{x},t)$$

$$=-i\int\frac{d^3k}{(2\pi)^3}\sqrt{\frac{\hbar\omega_k}{2}}\left\{a(\vec{k})e^{i(\vec{k}\cdot x-\omega_k t)}-a^\dagger(\vec{k})e^{-i(\vec{k}\cdot\vec{x}-\omega_k t)}\right\}$$

Finally, it is also useful to rewrite $\phi_1(\vec{k})=\frac{1}{2}\left[\tilde{\phi}(\vec{k},t)+\frac{i}{\omega_p}\tilde{\pi}(\vec{k},t)\right]e^{i\omega_p t}$ $=\frac{1}{2}\int d^3xe^{-i(\vec{k}\cdot\vec{x}-\omega_p t)}\left[\phi(\vec{x},t)+\frac{i}{\omega_p}\pi(\vec{x},t)\right]$ in terms of the properly normalized annihilation operator, so that one can express the annihilation operator in terms of the field and canonical momentum density:

$$a(\vec{k})=\sqrt{\frac{\omega_k}{2\hbar}}\int d^3xe^{-i(\vec{k}\cdot\vec{x}-\omega_p t)}\left[\phi(\vec{x},t)-\frac{i}{\omega_p}\pi(\vec{x},t)\right]$$

We might also want the corresponding formula for the creation operator, which can be found simply by taking the adjoint of the above equation:

$$a^\dagger(\vec{k})=\sqrt{\frac{\omega_k}{2\hbar}}\int d^3xe^{-i(\vec{k}\cdot\vec{x}-\omega_p t)}\left[\phi(\vec{x},t)-\frac{i}{\omega_p}\pi(\vec{x},t)\right]$$

Euler-Maclaurin Summation Formula

The Euler-Maclaurin formula relates the sum of a function evaluated at evenly spaced points to the corresponding integral approximation, providing a systematic method of calculating corrections in terms of the derivatives of the function evaluated at the end-points. Consider first a function defined on the interval $-1 \le x \le 1$, for which we can imagine approximating the sum of f(−1) + f(1) by the integral of the function over the interval:

$$f(-1)+f(1)=\int_{-1}^{1} dx\, f(x)+R_1$$

where R_1 represents a correction term that we want to understand. One can find an exact expression for R_1 by applying an integration by parts to the integral:

$$\int_{-1}^{1} dx\, f(x)=f(-1)+f(1)-\int_{-1}^{1} dx\, x f'(x)$$

So,

$$R_1=\int_{-1}^{1} dx\, x f'(x)$$

Where, a prime denotes a derivative with respect to x.

Expansion by Successive Integrations by Parts

We want an approximation that is useful for smooth functions f(x), and a smooth function is one for which the higher derivatives tend to be small. Therefore, if we can extract more terms in a way that leaves a remainder term that depends only on high derivatives of the function, then we have made progress. This can be accomplished by successively integrating by parts, each time differentiating f(x) and integrating the function that multiplies it. We can define a set of functions:

$$V_0(x)\equiv 1,\quad V_1(x)\equiv x$$

And,

$$V_n(x)\equiv\int dx\, V_{n-1}(x)$$

$V_n(x)\equiv\int dx\, V_{n-1}(x)$ is not quite well-defined, however, because each indefinite integral is defined only up to an arbitrary constant of integration. Regardless of how these constants of integration are chosen, however, one can rewrite $f(-1)+f(1)=\int_{-1}^{1} dx\, f(x)+R_1$ by using $R_1=\int_{-1}^{1} dx\, x f'(x)$, and then successively integrating by parts:

$$f(-1)+f(1)=\int_{-1}^{1} dx\, f(x)+\int_{-1}^{1} dx V_1(x)\, f'(x)$$

$$=\int_{-1}^{1}dx(x)+V_2(1)f'(1)-V_2(-1)f'(-1)-\int_{-1}^{1}dxV_2(x)f''(x)$$

$$=\int dxf(x)+\left[V\ (1)f'(1)f'(-1)\right]$$

$$=\left[V_3(1)f''(1)-V_3(-1)f''(-1)\right]+\left[V_4(1)f'''(1)-V_4(-1)f'''(-1)\right]+....$$

$$+\left[V_{2n}(1)f^{2n-1}(1)-V_{2n}(-1)f^{2n-1}(-1)\right]-\int\ dxV_{2n}(x)f^{2n}(x)$$

$$=\int_{-1}^{1}dxf(x)+\sum_{1\ 2}^{2n}(-1)\ \left[V_1(1)f^{l-1}(1)-V_1(-1)f^{l-1}(-1)\right]-\int_{-1}^{1}dxV_{2n}(x)f^{2n}(x)$$

Where $f^{n}(x)$ denotes the nth derivative of f with respect to x.

Elimination of the Odd Contributions

Above equation is valid for any choice of integration constants in $V_n(x)\equiv\int dx\,V_{n-1}(x)$, so we can seek a choice that simplifies the result. $V_1(x)$ is odd under $x\rightarrow -x$. We can therefore choose the integration constants so that:

$$V_n(-x)=\begin{cases}V_n(x)\ \text{if n is even.}\\-V_n(x)\ \text{if n is odd}\end{cases}$$

This even/odd requirement uniquely fixes the integration constant in $V_n(x)\equiv\int dx\,V_{n-1}(x)$ when n is odd, because the sum of an odd function and a constant would no longer be odd. We are still free, however, to choose the integration constants when n is even. Using the even/odd property,

$$f(-1)+f(1)=\int_{-1}^{1}dxf(x)+\sum_{l=2}^{2n}(-1)^{l}\left[V_l(1)f^{l-1}(1)-V_l(-1)f^{l-1}(-1)\right]-\int_{-1}^{1}dxV_{2n}(x)f^{2n}(x)$$

The terms in V(1) for even ℓ involve the difference of $f^{\ell-1}$ at the two endpoints, while the terms for odd ℓ involve the sum.

$$f(-1)+f(1)=\int_{-1}^{1}dxf(x)+\sum_{l=2}^{2n}(-1)^{l}\left[V_l(1)f^{l-1}(1)-V_l(-1)f^{l-1}(-1)\right]-\int_{-1}^{1}dx\,V_{2n}(x)f^{2n}(x)$$

describes a single interval, however, and our goal is to obtain a formula valid for any number of intervals. We will do this by first generalizing previous equation to apply to an arbitrary interval $a\le x\le a+h$, and then applying it to each interval in a succession of evenly spaced intervals. When this succession is summed, the even ℓ terms involving the differences of the endpoints will cancel at each interior point, but the odd ℓ terms will add .The odd ℓ terms can therefore make a considerably more complicated contribution to the answer, but we can force them to vanish by using the remaining freedom in the choice of integration constants. When n is even in $V_n(x)\equiv\int dx\,V_{n-1}(x)$ we choose the integration constant so that, $\int_{-1}^{1}dxV_n(x)\equiv 0$.

$\int_{-1}^{1} dxV_n(x) \equiv 0$ is always true for odd functions, so it is true for all n > 0. It then follows for all n > 1 that:

$$V_n(1) - V_n(-1) = \int_{-1}^{1} dxV_n(x) \equiv 0$$

If n is odd then $V_n(-x) = \begin{cases} V_n(x) & \text{if n is even.} \\ -V_n(x) & \text{if n is odd} \end{cases}$ implies that $V_n(-1) = -V_n(1)$, and so,

$V_n(1) = V_n(-1) = 0$ for all odd $n > 1$ as desired.

The V_n's are now uniquely defined. In our construction we used the anti-symmetry property to fix the constant of integration for odd n, and the vanishing of the integral to fix the integration constant for even n. However, holds also for odd n, and is sufficient to fix the integration constant for the odd n cases. The functions $V_n(x)$ can therefore be defined succinctly by,

$$V_0(x) \equiv 1,$$
$$V_n(x) \equiv \int dx\, V_{n-1}(x) \text{ and}$$
$$\int_{-1}^{1} dxV_n(x) \equiv 0 \quad (\text{for } n > 0)$$

We can use these properties to build a table for the lowest values of n:

n	$V_n(x)$	$V_n(1)$
0	$V_0(x) = 1$	1
1	$V_1(x) = x$	1
2	$V_2(x) = \frac{x^2}{2} - \frac{1}{6}$	$\frac{1}{3}$
3	$V_3(x) = \frac{x^3}{6} - \frac{1}{6}$	0
4	$V_4(x) = \frac{x^4}{24} - \frac{x^2}{12} + \frac{7}{360}$	$-\frac{1}{45}$
5	$V_5(x) = \frac{x^5}{120} - \frac{x^3}{36} + \frac{7x}{360}$	0
6	$V_6(x) = \frac{x^6}{720} - \frac{x^4}{144} + \frac{7x^2}{720} - \frac{31}{15120}$	$\frac{2}{945}$

$V_n(1) = V_n(-1) = 0$ for all odd $n > 1$ as desired guarantees that only the even terms contribute to equation, so we can set ℓ = 2k and rewrite,

$$f(-1)+f(1)=\int_{-1}^{1} dx f(x)+\sum_{k=1}^{n} V_{2k}(1)\left[f^{2k-1}(1)-f^{2k-1}(-1)\right]$$

$$-\int_{-1}^{1} dx V_{2n}(x) f^{2n}(x)$$

Application to an Arbitrary Interval

To apply previous equation to an arbitrary interval a < x < a + h, one needs only to change variables. Let $f(x) = \tilde{f}(\tilde{x})$, where,

$$\tilde{x} = (x+1)\frac{h}{2}+a$$

Rewriting $f(-1)+f(1)=\int_{-1}^{1} dx f(x)+\sum_{k=1}^{n} V_{2k}(1)\left[f^{2k-1}(1)-f^{2k-1}(-1)\right]-\int_{-1}^{1} dx V_{2n}(x) f^{2n}(x)$

in terms of), while dividing the whole equation by 2 for later convenience, one has,

$$\frac{1}{2}\left[\tilde{f}(a)+\tilde{f}(a+h)\right]=\frac{1}{h}\int_{a}^{a+h} d\tilde{x}\, \tilde{f}(\tilde{x})$$

$$+\frac{1}{2}\sum_{k=1}^{n}\left(\frac{h}{2}\right)^{2k-1} V_{2k}(1)\left[\tilde{f}^{2k-1}(a+h)-\tilde{f}^{2k-1}(a)\right]$$

$$-\frac{1}{2}\left(\frac{h}{2}\right)^{2n-1}\int_{a}^{a+h} d\tilde{x}\, V_{2n}\left(\frac{2}{h}(\tilde{x}-a)-1\right)\tilde{f}^{2n}(\tilde{x})$$

where $\tilde{f}^n(\tilde{x})$ denotes the nth derivative of $\tilde{f}$ with respect to its argument $\tilde{x}$. Now that the original f and x have been eliminated, we can drop the tilde superscripts that appear in above equation.

Application to an Arbitrary Sum of Intervals

The problem can now be completed by extending above equation to an interval a < x < b, divided into m steps of size h = (b–a)/m. Adding an expression of the form for each interval of size h, one has,

$$\sum_{k=0}^{m} f(a+kh)=\frac{1}{h}\int_{a}^{b} dx\, f(x)+\frac{1}{2}\left[f(a)+f(b)\right]$$

$$+\frac{1}{2}\sum_{k=1}^{n}\left(\frac{h}{2}\right)^{2k-1} V_{2k}(1)\left[f^{2k-1}(b)-f^{2k-1}(a)\right]$$

$$-\frac{1}{2}\left(\frac{h}{2}\right)^{2n-1}\int_{a}^{a+h} dx\, V_{2n}\left(\frac{2}{h}(x-a)-1\right)\sum_{k=0}^{m-1} f^{2n}(x+kh)$$

When one adds up the left-hand sides of the expressions of the form, all the terms have coefficient 1 except for the first and last term, each of which have coefficient $\frac{1}{2}$. In above equation, the sum is written with all terms having coefficient 1, and the correction for the first and last term has been moved to the right-hand side. For all practical purposes, above equation is all that is necessary. For the Casimir application a = 0, b = ∞, and h = 1, so above equation becomes,

$$\sum_{k=0}^{\infty} f(k) = \int_0^{\infty} dx\, f(x) + \frac{1}{2} f(0) - \frac{1}{12} f'(0) + \frac{1}{720} f'''(0) - \frac{1}{30{,}240} f'''''(0) + \,..$$

Connection to the Bernoulli Numbers

Let F(x) have its first 2n derivatives continuous on an interval (a, b). Divide the interval into m equal parts and let h = (b – a)/m.Then for some θ, 1 > θ > 0, depending on $F^{(2n)}(x)$ on (a, b), we have,

$$\sum_{k=0}^{m} F(a+kh) = \frac{1}{h}\int_a^b F(t)\,dt + \frac{1}{2}\{F(b)+F(a)\}$$

$$+\sum_{k=1}^{n-1} \frac{h^{2k-1}}{(2k)!} B_{2k} \left\{F^{(2k-1)}(b) - F^{(2k-1)}(a)\right\}$$

$$+\frac{h^{2n}}{(2n)!} B_{2n} \sum_{k=0}^{m-1} F^{(2n)}(a+kh+\theta h).$$

Disregarding for now the remainder term (the 3rd term on the right), above equation agreeswith $\sum_{k=0}^{m} f(a+kh) = \frac{1}{h}\int_a^b dx\, f(x) + \frac{1}{2}[f(a)+f(b)] + \frac{1}{2}\sum_{k=1}^{n}\left(\frac{h}{2}\right)^{2k-1} V_{2k}(1)\left[f^{2k-1}(b) - f^{2k-1}(a)\right]$

$-\frac{1}{2}\left(\frac{h}{2}\right)^{2n-1} \int_a^{a+h} dx\, V_{2n}\left(\frac{2}{h}(x-a)-1\right) \sum_{k=0}^{m-1} f^{2n}(x+kh)$ provided that,

$$\frac{B_{2k}}{(2k)!} = \frac{V_{2k}(1)}{2^{2k}}.$$

The Bernoulli numbers B_n are defined in terms of the Bernoulli polynomials $B_n(x)$, which are defined by the generating function,

$$\frac{te^{xt}}{e^t - 1} = \sum_{n=0}^{\infty} B_n(x) \frac{t^n}{n!} \quad (\text{for } t<2\pi).$$

The Bernoulli numbers are given by,

$$B_n = B_n(0).$$

From these relations one can easily calculate the lowest Bernoulli polynomials:

n	$B_n(x)$	$B_n = B_n(0)$
0	$B_0(x) = 1$	1
1	$B_1(x) = x - \frac{1}{2}$	$-\frac{1}{2}$
2	$B_2(x) = x^2 - x + \frac{1}{6}$	$\frac{1}{6}$
3	$B_3(x) = x^3 - \frac{3}{2}x^2 + \frac{1}{2}x$	0
4	$B_4(x) = x^4 - 2x^3 + x^2 - \frac{1}{30}$	$-\frac{1}{30}$
5	$B_5(x) = x^5 - \frac{5}{2}x^4 + \frac{5}{3}x^3 - \frac{1}{6}x$	0
6	$B_6(x) = x^6 - 3x^5 + \frac{5}{2}x^4 - \frac{1}{2}x^2 + \frac{1}{42}$	$\frac{1}{42}$

By comparing above equations with $V_n(x)\ldots \frac{2}{945}$, one can conjecture the equality,

$$V_n(x) = \tilde{V}_n(x),$$

Where,

$$\tilde{V}_n(x) \equiv \frac{2^n}{n!} B_n\left(\frac{x+1}{2}\right).$$

To prove this equality, it is sufficient to verify that $\tilde{V}_n(x)$ satisfies the relations $V_0(x) \equiv 1, V_n(x) \equiv \int dx\, V_{n-1}(x)$ and $\int_{-1}^{1} dx V_n(x) \equiv 0$ (for $n > 0$), since these were the relations that defined $V_n(x)$. It is straightforward to determine the properties required for $B_n(x)$ so that $\tilde{V}_n(x)$ obeys each of the relations $V_0(x) \equiv 1, V_n(x) \equiv \int dx\, V_{n-1}(x)$ and and $\int_{-1}^{1} dx V_n(x) \equiv 0$ (for $n > 0$):

$$B_0(x) = 1$$

$$B_n(x) = n \int dx\, B_{n-1}(x), \text{ and}$$

$$\int_0^1 dx\, B_n(x) = 0 \qquad (\text{for } n > 0).$$

$B_0(x) = 1$ was already written within $B_n(x) .... \frac{1}{42}$. To verify $B_n(x) = n\int dx\, B_{n-1}(x)$, differentiate the generating equation $\frac{te^{xt}}{e^t - 1} = \sum_{n=0}^{\infty} B_n(x)\frac{t^n}{n!}$ (for $t < 2\pi$) with respect to x:

$$\frac{t^2\, e^{xt}}{e^t - 1} = \sum_{n=0}^{\infty} \frac{dB_n(x)}{dx}\frac{t^n}{n!}.$$

One can obtain another expansion of the same quantity by multiplying the generating equation by t:

$$\frac{t^2\, e^{xt}}{e^t - 1} = \sum_{n=0}^{\infty} B_n(x)\frac{t^{n+1}}{n!}$$
$$= \sum_{n=0}^{\infty} B_{n-1}\frac{t^n}{(n-1)!}.$$

Comparing like powers of t in expansions $\frac{t^2\, e^{xt}}{e^t - 1} = \sum_{n=0}^{\infty} \frac{dB_n(x)}{dx}\frac{t^n}{n!}$ and $\frac{t^2\, e^{xt}}{e^t - 1} = \sum_{n=0}^{\infty} B_n(x)\frac{t^{n+1}}{n!} = \sum_{n=0}^{\infty} B_{n-1}\frac{t^n}{(n-1)!}$, one finds,

$\frac{dB_n}{dx} = n\, B_{n-1}(x)$, which is equivalent to $B_n(x) = n\int dx\, B_{n-1}(x)$. Finally, to verify $\int_0^1 dx\, B_n(x) = 0$, one can integrate the generating equation over x from 0 to 1:

$$\int_0^1 dx \frac{te^{xt}}{e^t - 1} = 1 = \sum_{n=0}^{\infty} \frac{t^n}{n!} \int_0^1 dx\, B_n(x).$$

Again by comparing powers of t, one verifies $\int_0^1 dx\, B_n(x) = 0$. Thus $\tilde{V}_n(x)$ obeys all of the relations $V_0(x) \equiv 1, V_n(x) \equiv \int dx\, V_{n-1}(x)$ and $\int_{-1}^1 dx V_n(x) \equiv 0$ (for $n > 0$), and hence $V_n(x) = \tilde{V}_n(x)$. Having established $V_n(x) = \tilde{V}_n(x)$, it follows immediately that, $V_{2k}(1) = V_{2k}(-1) = \frac{2^{2k}}{(2k)!} B_{2k}(0)$, which is just what is needed to verify $\frac{B_{2k}}{(2k)!} = \frac{V_{2k}(1)}{2^{2k}}$, and hence the agreement of our series expansion with that of A&S.

Distributions and the Fourier Transform

In QFT it is common to encounter integrals that are not well-defined. Researchers, for example, give the following formula for the two point function $\langle o|\phi(x)\,\phi(y)|o\rangle$ for spacelike separations $(x-y)^2 = -r^2$:

$$D(r) = \frac{-i}{2(2\pi)^2 r}\int_{-\infty}^{\infty} dp \frac{pe^{ipr}}{\sqrt{p^2+m^2}}.$$

If this integral is defined in the usual way as,

$$\lim_{\Lambda\to\infty}\int_{-\Lambda}^{\Lambda} dp \frac{pe^{ipr}}{\sqrt{p^2+m^2}},$$

then it does not exist. The integral can be defined by putting in a convergence factor $e^{-\epsilon|p|}$:

$$\lim_{\epsilon\to 0}\int_{-\infty}^{\infty} dp \frac{pe^{ipr}\,e^{-\epsilon|p|}}{\sqrt{p^2+m^2}}.$$

But how does one know whether a different convergence factor would get the same result? One way to resolve these issues is to treat the ambiguous quantity as a distribution, rather than a function. All tempered distributions have Fourier transforms, which are also tempered distributions.

1. Distribution: A distribution is a linear mapping from a space of test functions to real or complex numbers. (An operator-valued distribution maps test functions into operators).

2. Test Functions: The space of test functions $\{\varphi(t)\}$ determines what type of distribution one is discussing. The test functions for tempered distributions belong to "Schwartz space," the space of functions which are infinitely differentiable, and the function and each of its derivatives fall off faster than any power for large t. The Gaussian is a good example of a Schwartz function. Any function in Schwartz space has a Fourier transform in Schwartz space.(The Fourier transform of a Gaussian is a Gaussian).

3. Functions as Distributions: Distributions are sometimes called generalized functions, which suggests that a function is also a distribution. This is not quite true, but a wide range of functions can also be thought of as distributions. Given any function $f(t)$ which is piecewise continuous and bounded by some power of t for large t, one can define a corresponding distribution T_f by,

$$T_f[\varphi] \equiv \int_{-\infty}^{\infty} dt\, \phi(t)\, \varphi(t).$$

Since $\varphi(t)$ falls off faster than any power, this integral will converge because the class of $\varphi(t)$'s is very restricted, the class of possible $f(t)$'s is very large.

Fourier Transform: For any function $f(t)$ which is integrable, meaning that, $\int_{-\infty}^{\infty} dt\,|f(t)|$ converges, define, $\tilde{f}(\omega) \equiv \int_{-\infty}^{\infty} dt\, e^{-i\omega t} f(t)$.

4. Fourier Transform of a Distribution: To motivate the definition, suppose $f(t)$ is integrable, and consider,

$$\begin{aligned} T_{\tilde{f}}[\varphi] &= \int_{-\infty}^{\infty} \tilde{f}(\omega)\,\varphi(\omega)\,d\omega \\ &= \int_{-\infty}^{\infty} dt \int_{-\infty}^{\infty} e^{-i\omega t} f(t)\,\varphi(\omega)\,d\omega \\ &= \int_{-\infty}^{\infty} dt\, f(t)\,\tilde{\varphi}(t) \\ &= T_f[\tilde{\varphi}]. \end{aligned}$$

These integrals are absolutely convergent, so there is no problem about interchanging the order of integration. So, for any distribution T, define its Fourier transform by, $\tilde{T}[\varphi] \equiv T[\tilde{\varphi}]$.

Any function f(t) which is piecewise continuous and bounded by some power of t for large t can define a distribution, and can therefore be Fourier transformed as a distribution.

5. Relation to ϵ convergence factor: Suppose f(t) is not integrable, and so does not have a Fourier transform. Suppose, however, that there exists a continuous sequence of "regulated functions" $f_\epsilon(t)$ which are integrable for $\epsilon > 0$, which satisfy,

$|f_\epsilon(t)| < |f(t)|$, and which for each t satisfy, $\lim_{\epsilon \to 0} f_\epsilon(t) = f(t)$.

Example: $f_\epsilon(t) = f(t)\,e^{-\epsilon|t|}$. The regulator that we used for the two-point function at spacelike separations has this property. To show: if we Fourier transform $f_\epsilon(t)$ and take the limit $\epsilon \to 0$ at the end, it is the same as the distribution-theory definition of the Fourier transform.

Proof: The distribution-theory definition of the Fourier transform is,

$$\begin{aligned} \tilde{T}[\varphi] &\equiv T[\tilde{\varphi}] \\ &= \int_{-\infty}^{\infty} dt\, f(t)\,\tilde{\varphi}(t). \end{aligned}$$

The ϵ prescription is to use, $T_f^*[\varphi] \equiv \lim_{\epsilon \to 0} T_{\tilde{f}_\epsilon}[\varphi]$. We need to show these are equivalent. Use,

$$\begin{aligned} T_f^*[\varphi] &= \lim_{\epsilon \to 0} \int_{-\infty}^{\infty} d\omega\, \tilde{f}_\epsilon(\omega)\,\varphi(\omega) \\ &= \lim_{\epsilon \to 0} \int_{-\infty}^{\infty} d\omega \int_{-\infty}^{\infty} dt\, e^{-i\omega t} f_\epsilon(t)\,\varphi(\omega) \\ &= \lim_{\epsilon \to 0} \int_{-\infty}^{\infty} dt\, f_\epsilon(t)\,\tilde{\varphi}(t). \end{aligned}$$

Last step is proven with Lebesgue's Dominated Convergence Theorem: If $h_\in(t)$ is a sequence of functions for which,

$$\lim_{\in\to0} h_\in(t) = h(t) \qquad \text{for all t,}$$

and if there exists a function g(t) for which, $\int dt\, g(t)$ converges, and for which, $g(t) \geq |h_\in(t)|$ for all t and all $\in$, then, $\lim_{\in\to0}\int dt\, h_\in(t) = \int dt\, h(t)$.

By the way, that the existence of the integrable bounding function g(x) is absolutely necessary. A simple example of a function $h_\in(t)$ for which one cannot bring the limit through the integral sign would be a function that looks something like:

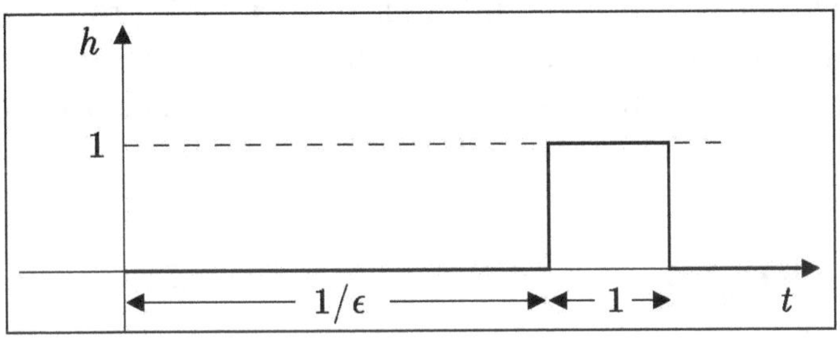

Analytically, this function can be written as,

$$h_\in(t) = \begin{cases} 1 & \text{if } \frac{1}{\in} < t < \frac{1}{\in} + 1 \\ 0 & \text{otherwise.} \end{cases}$$

The square well moves infinitely far to the right as $\in \to 0$, so $h_\in(t) \to 0$ for any t. But the integral of the curve is 1 for any $\in$, and hence it is 1 in the limit. The Lebesgue Dominated Convergence theorem excludes functions like this, because any bounding function g(t) must be ≥ 1 for all t, so g(t) cannot be integrable. The theorem does apply, however, to,

$$\lim_{\in\to0}\int_{-\infty}^{\infty} dt\, f_\in(t)\,\tilde{\varphi}(t).$$

Take,

$$h_\in(t) = f_\in(t)\,\tilde{\varphi}(t),$$
$$h(t) = f(t)\,\tilde{\varphi}(t),$$

and, $g(t) = |f(t)\,\tilde{\varphi}(t)|$.

6. Bottom Line: The prescription used by physicists is equivalent to the unambiguous definition of the Fourier transform in tempered-distribution theory. That is, if the function to be Fourier-transformed f(t) is not integrable, one can proceed as long

as one can find an integrable regulator $f_{\in}(t)$ such that, $|f_{\in}(t)| < |f(t)|$, and for each t, $\lim_{\in \to 0} f_{\in}(t) = f(t)$.

One can then Fourier transform $f_{\in}(t)$ instead. In the general case one cannot take the limit $\in \to 0$ immediately, but one must leave in the expression for the distribution. Only after the distribution is evaluated for a particular test function can the limit $\in \to 0$ be taken. Remember, for example, that we wrote the Fourier transform of the Feynman propagator as,

$$\frac{i}{p^2 - m^2 + i\in}.$$

With the $\in$ in place one can carry out integrals involving the propagator, and then one can take the limit $\in \to 0$ at the end. If one tried to set the $\in$ term to zero immediately, then the poles in the propagator would lead to ill-defined integrations.

String Theory

String theory in particle physics is a theory that attempts to merge quantum mechanics with Albert Einstein's general theory of relativity. The name string theory comes from the modeling of subatomic particles as tiny one-dimensional "stringlike" entities rather than the more conventional approach in which they are modeled as zero-dimensional point particles. The theory envisions that a string undergoing a particular mode of vibration corresponds to a particle with definite properties such as mass and charge. In the 1980s, physicists realized that string theory had the potential to incorporate all four of nature's forces, gravity, electromagnetism, strong force, and weak force and all types of matter in a single quantum mechanical framework, suggesting that it might be the long-sought unified field theory. While string theory is still a vibrant area of research that is undergoing rapid development, it remains primarily a mathematical construct because it has yet to make contact with experimental observations.

The Bosonic String

It is well known that the Lagrangian of a relativistic point particle is proportional to the corresponding invariant length, the world-line described by the particle while moving in the space-time. Our first attempt to describe the dynamics of a string is to construct its Lagrangian, which based on the relativistic particle case, we expect it to be proportional to the invariant area described by the string as it moves in a D-dimensional space-time. This is called the string world-sheet and it is a 2-dimensional object with an induced metric denoted by $G_{\alpha\beta}$. The action is given by,

$$S = -T\int d^2\sigma \sqrt{-\det(G_{\alpha\beta})},$$

where the world-sheet is parametrized by $\sigma_1 = \sigma$ (a space-like coordinate), and $\sigma_0 = \tau$ (a time-like coordinate). After some direct calculations, the above action can be written in terms of the D-dimensional coordinates X^M which characterize the position of the world-sheet in the space-time. This is the Nambu-Goto action and it is given by,

$$S_{NG} = -T \int d\sigma\, d\tau \sqrt{\left(\dot{X}\cdot X'\right)^2 + \dot{X}^2 X'^2},$$

Where,

$$\dot{X}^M = \frac{\partial X^M}{\partial \tau} \quad X'^M = \frac{\partial X^M}{\partial \sigma}.$$

However this action is difficult to quantize. An easiest way to do it requires an action without the square root of the field derivatives. The desired action is known as the Polyakov action and it is equivalent to the Nambu-Goto action at the classical level in the sense that it gives rise to the same equations of motion for the fields X^M. The Polyakov action is, $S_P = -\frac{1}{2}\int d^2\sigma \sqrt{-h}\, h_{\alpha\beta}\, \partial_\alpha X \cdot \partial_\beta X,$ where $h = \det(h_{\alpha\beta}), h_{\alpha\beta}$ is the induced metric on the world-sheet, and α and β runs on 0 and 1. This action posses the following symmetries.

- Space-time Poincaré symmetry: The action is invariant under Poincaré transformations. It is a global symmetry.
- Reparametrization: This is non-trivial gauge symmetry and represent a large extra symmetry on the world sheet. Essentially tells us that we can select different coordinate frames on the world-sheet.
- Weyl transformation: This symmetry represents the invariance of the action under rescaling of the metric. By transforming the induced metric $h_{\alpha\beta}$ to $\Omega^2 h_{\alpha\beta}$ the equations of motion are the same.

Since we have three constraints given by these symmetries, it is possible to fix the corresponding three parameters in $h_{\alpha\beta}$. By reparametrization invariance we can choose two components of hαβ, such that only one component remains free. But Weyl invariance fixes it. Therefore, the induced metric hαβ is fixed as $h_{\alpha\beta} = \eta_{\alpha\beta}$, where $\eta_{\alpha\beta}$ is the flat 2-dimensional Minkowskian metric. With this gauge, the Polyakov action reduces to,

$$S = \frac{T}{2}\int d^2\sigma\left(\dot{X}^2 \cdot X'^2\right).$$

The corresponding equation of motion with respect to the fields X^M, up to some boundary conditions, are given by, $\partial_\alpha \partial^\alpha X = 0$, which is a differential wave equation. Their solutions are then a linear combination of linear terms in τ and σ and periodic functions.

Such solutions represents undulating modes of the string. If some of the massless fields have the properties of already known fields in the context of quantum field theory, quantum oscillations of a string produces the particles we are familiar with. For instance, let us say that we find that one of the modes is massless, transforms under the Lorentz group as a vector boson, has a U(1) internal symmetry and has a spin 1. Such mode is indeed a photon. Things are nevertheless, a little bit more complicated, but we shall see actually that some modes have all the desired properties with respect to the particles we already know. Fixing the gauge as before, implies that each component of the energy-momentum tensor vanishes, $T_{\alpha\beta} = \partial_\alpha X \partial_\beta X - \frac{1}{2} h_{\alpha\beta} h^{\gamma\delta} \partial_\gamma X \partial_\delta X = 0$, and by using the flat induced metric, the components read,

$$T_{\alpha\beta} \doteq \begin{bmatrix} \frac{1}{2}\left(\dot{X}^2 \cdot X'^2\right) & \dot{X}X' \\ \dot{X}X' & \frac{1}{2}\left(\dot{X}^2 \cdot X'^2\right) \end{bmatrix} = 0 .$$

This expression has a geometrical interpretation. The gauge selection implies that the coordinate framework on the world-sheet is orthogonal. This allows us to choose a particular useful system of coordinates: the light-cone coordinates.

Boundary Conditions and D-branes

The stationary points are chosen by demanding the invariance of the action under the shifts $X^M \to X^M + \delta X^M$. The variation of the action reads, $\delta S = \frac{T}{2} \int d^2\sigma \left(\partial_\alpha \partial^\alpha X_M\right) \delta X^M - T \int d\tau \left[X'_M \delta X^M \big|_{\sigma=\pi} - X'_M \delta X^M \big|_{\sigma=0} \right]$, from which, the vanishing of the boundary terms leads to different conditions on the string. Strings which endpoints are identified, are called closed strings, while open strings presents two variations.

Table: Boundary conditions on the bosonic string.

Type	Boundary Condition
Closed	$X^M(\sigma,\tau) = X^M(\sigma+\pi,\tau)$
Open/Neumann	$X'_M = 0,\ \sigma = 0,\pi$
Open/Dirichlet	$\delta X^M = 0,\ \sigma = 0,\pi$

For Neumann boundary conditions there is not momentum flowing away from the string endpoints. These strings are free to move over the whole D-dimensional space-time, meaning that we can impose Nuemann conditions on all field coordinates X^M. We can also consider an open string with all its field coordinates fulfilling Dirichlet boundary conditions. Each of the endpoints of the strings, are fixed on a point in a

D-dimensional space-time. Notice as well that such point is localized in time. This is indeed an instanton of the theory. However, let us consider the richest case, in which we impose Dirichlet boundary conditions on some coordinates and Neumann on others. Let us say that for M = 0, 1, . . . , p, X^M satisfies Neumann boundary conditions, while for M = p + 1, . . . , d, it satisfies Dirichlet conditions. Henceforth, the endpoints of the string will be attached to move freely on a p + 1-dimensional hyperplane. Notice that they cannot leave this object unless their endpoints coincide. If this happens, the open string becomes closed and it is free to escape away from the D-brane. In this sense we say that D-branes emit closed strings.

It turns out that such objects are very important in string theory. They are named Dp-branes: p- dimensional branes (generalization of membranes) on which the endpoints of an open string are attached to. Orthogonal coordinates of the endpoints satisfy Dirichlet boundary conditions (the "D" of D-brane refers to Dirtichlet boundary conditions). For the bosonic case, it is possible to have different Dp-branes of all dimensionalities, from p = −1 to p = d.

Solution to the Equations of Motion

We are going to solve the equations of motion in a particular coordinate-frame on the string world-sheet, known as the light-cone coordinates. The advantages of this selection are that we shall arrive to the solutions in a faster and easier way. The price to pay is however, that we have lost covariance in our description of the string dynamics. Also, by selecting this frame, one can check that there are two unphysical degrees of freedom, corresponding to stretching and rotating a string. In a covariant formalism, one is able to arrive at the same conclusion, but in a richer way, i.e., by analyzing the anomalies, ghosts fields and by using the powerful conformal field theory in the string world-sheet. The light-cone coordinates are given by $\sigma^{\pm} = \tau \pm \sigma$, from which the equations of motion read $\partial_+ \partial_- X^M = 0$.

For the closed string, the corresponding solution can be written in terms of two waves by, $X^M(\sigma, \tau) = X_R^M(\sigma^-) + X_L^M(\sigma^+)$, which represent left and right movers. The most general solution for the closed string given by,

$$X_R^M = \frac{1}{2}x^M + \alpha' p^M \sigma^- + \frac{i}{2} l_s \sum \frac{1}{n} \alpha_n^M e^{-2in\sigma^-} \text{ and } X_L^M = \frac{1}{2}x^M + \alpha' p^M \sigma^+ + \frac{i}{2} l_s \sum \frac{1}{n} \tilde{\alpha}_n^M e^{-2in\sigma^+}.$$

Since $X^M(\sigma, \tau)$ is required to be real, the left and right movers X_L^M and X_R^M must be real function as well. This implies that x^M (center of mass) and p^M (momentum of the center of mass) are also real and in consequence, the modes of the string must obey that $\alpha_{-n}^M = (\alpha_n^M)^*, \tilde{\alpha}_{-n}^M = (\tilde{\alpha}_n^M)^*$ for all n and where we have used $\alpha^M = \sqrt{\alpha'/2}\, p^M$. Introducing X_L and X_R into the energy-momentum tensor we obtain that, $T_{--} = \alpha' \sum L_m e^{-2im\sigma^-}$, $T_{++} = \alpha' \sum \tilde{L}_m e^{-2im\sigma^+}$, where the Fourier coefficients are actually the Virasoro generators given by, $L_m = \frac{1}{2} \sum \alpha_{m-n}\, \alpha_n$ $\tilde{L}_m = \frac{1}{2} \sum \tilde{\alpha}_{m-n}\, \tilde{\alpha}_n$.

So the requirement of $T_{--} = T_{++} = 0$ is reflected in the fact that all the modes must vanish. This constitutes a classical constraint for all modes, $L_m = \tilde{L}_m = 0$. An interesting and special case concerns the zero mode, since it offers a way to compute the mass associated to different Fourier modes. Hence, for m = 0,

$$\begin{aligned} L_0 + \tilde{L}_0 &= \frac{1}{2}\left(\alpha_0^2 + \alpha_0^2\right) + \sum_{n \neq 0} \left(\alpha_{-n} \cdot \alpha_n, \tilde{\alpha}_{-n} \cdot \tilde{\alpha}_n\right) \\ &= \alpha' p^M p_M + \sum_{n \neq 0} \left(\alpha_{-n} \cdot \alpha_n, \tilde{\alpha}_{-n} \cdot \tilde{\alpha}_n\right) \end{aligned}$$

Since in a relativistic theory, $p^2 = -m^2$, we arrive at the mass expression for the closed string modes, $M^2 = \frac{1}{\alpha} \sum_{n=1}^{\infty} \alpha_{-n} \cdot \alpha_n$, where the most general solution of the equations of motion is given by,

$$x^M(\sigma, \tau) = x^M + \frac{\alpha'}{2} p^M \tau + \frac{i\alpha'}{2} \sum \frac{1}{m} \alpha_n^M e^{-im\tau} \cos(m\sigma).$$

Canonical Quantization

There are many ways to quantize the string, being the most popular the covariant quantization. However, since we want to study the quantum string features in a short way, we shall take the canonical quantization procedure. Essentially, this consists in promoting the Fourier modes α_n to operators which satisfy the algebra,

$$\left[\alpha_m^M, \alpha_n^N\right] = \left[\tilde{\alpha}_m^M, \tilde{\alpha}_n^N\right] = \eta^{MN} \delta_{m+m,0}, \left[\alpha_m^M, \tilde{\alpha}_n^N\right] = 0.$$

This is the algebra that satisfies every rising and annihilator operator in the usual quantum harmonic oscillator, meaning that each quantum mode, now promoted to be an operator, creates or destroys quantum states. However, since now the modes are non-commuting operators for m = 0, we must order the product appearing on the Virasoro generator operators. Therefore, it is possible to order the products $\alpha_{-n} \cdot \alpha_n$ (and similarly for left movers) up to the addition of a constant a. In this context, the mass term for the open string is written as,

$$M^2 = \frac{1}{\alpha'}\left(\sum \alpha_{-n} \cdot \alpha_n - a\right),$$

while for the closed string, it reads,

$$M^2 = \frac{2}{\alpha'}\left(\sum_{n=1}^{\infty} \left(\alpha_{-n} \cdot \alpha_n + \tilde{\alpha}_{-n} \cdot \tilde{\alpha}_n\right) - 2a\right),$$

where a is the zero point energy. Similarly as the quantum harmonic oscillator, the product of a lowering and a rising operators denotes another operator called the number operator N given by $N = \sum \alpha_{-n}\alpha_n$ for the left movers and $\tilde{N} = \sum \tilde{\alpha}_{-n}\tilde{\alpha}_n$ for the right ones. In this way, the normal ordening constant a cancels out for the difference $(L_o - \tilde{L}_o)|\phi\rangle = o$, which implies $N = \tilde{N}$. This is called the level-matching condition of the bosonic string.

The corresponding Hilbert space is constructed by acting with the modes α_n^i on a vacuum $|\Omega\rangle$ annihilated by the lowering operators α_{-n} and $\tilde{\alpha}_{-n}$. Although not explicitly shown in this notes, the use of the light-cone coordinates forces some coordinates to be non-dynamical. For that reason, the index i in the operators modes runs over D−2 values. Before constructing the quantum states, let us comment on the value of the ordering constant a, appearing in the mass expressions for the string. The value of such constant can be computed as follows. At first sight, it seems we are dwelling with a divergent quantity, since,

$$a = \frac{1}{2}(D-2)\sum_{n=1}^{\infty} n.$$

However this sum has an analytic continuation and it is computed by using the ζ-function, which is defined as,

$$\zeta(s) = \sum_{n=1}^{\infty} n^{-s}.$$

Table: Closed string spectrum. For D = 26, the second state is masless.

State	Level	M^2/α
$\|\Omega\rangle_L \otimes \|\Omega\rangle_R$	$N = \tilde{N} = o$	D-2
$\alpha_{-1}^i \tilde{\alpha}_{-1}^i \|\Omega\rangle_L \otimes \|\Omega\rangle_R$	$N = \tilde{N} = 1$	$\frac{26-D}{12}$

This complex function has as unique analytic continuation at s = −1, where it takes the value ζ(−1) = −1/12. Therefore by comparison we obtain that,

$$a = -\frac{D-2}{24}.$$

Quantum states are constructed by acting the rising operator modes and by restricting to the level match conditions. This means that for every left mover operator, there must be the corresponding right one. Therefore, there are an infinite tower of string states, generically given by the application of rising operators on the closed vacuum

$|\Omega\rangle_L \otimes |\Omega\rangle_R$ or on the open vacuum $|\Omega\rangle$. One way to construct gauge bosons, is by considering open strings.

The Bosonic String Spectrum

The first state corresponds to vanishing number operators, implying that such state is a tachyon, i.e., a quantum state with negative squared mass. Observe that this not depend on the value of D. The presence of tachyons, as it is now understood, reflects the fact that we are constructing a perturbative theory on a false vacuum. This means that there are some variable which must run down to a stable point on which the corresponding perturbative theory would not have tachyons in its spectrum. Closed string tachyons have been considered in literature, mainly in cosmological models.

The next state is given by a left and right number operators $N = \tilde{N} = 1$. Its mass is then (26–D)/12. It is a state with $(D-2)^2$ degrees of freedom. However, since the theory is Lorentz invariant, we expect that physical states belong to representations of the little group SO(D –2). The little group is the subgroup of the Lorentz group which leaves invariant the Ddimensional momentum of a particle. Therefore, only a massless state would have the necessary degrees of freedom of a little group representation (a massive one would transform with respect to the little group SO(D – 1)). This fixes completely the dimension number D to 26. Some comments are in order: first, the above argument, although not formal, yields the same result as the obtained by a covariant quantization. Second, by requiring the theory to be Lorentz invariant at the perturbative level, we have fixed the number of dimensions. These theories are called critical.

The critical dimension 26 can be as well obtained from the open string spectrum. The first excited state has (D – 2) degrees of freedom and must transform as a vector under the little group SO(D – 2). The number of degrees of freedom match only for a masless state, i.e., for D = 26.

Table: Open string spectrum for D = 26.

State	Level	M^2/α
$\|\Omega\rangle$	$N = 0$	-1
$\alpha^i_{-1}\|\Omega\rangle$	$N = 1$	0

Therefore, in the critical bosonic theory, the massless first excited state in the closed string, transforms as a two-index object under SO(24). It contains 24^2 degrees of freedom, which can be decomposed on irreducible representations of SO(24). Actually,

$$24^2 = 1 \oplus 276 \oplus 299,$$

corresponding to (all of them being massless) a scalar field φ called the dilaton, an antisymmetric two-index field B_{MN} and a symmetric two-index field G_{MN}. G_{MN} has all the properties we expect from a quantum particle of gravity, the graviton.

The Superstring

Bosonic string theory opened up a new and huge area of research. Quantum gravity, an old dream, was now a tangible subject. However, it failed by the first and the simplest feature we must have in any physical theory: the presence of fermions. Bosonic string theory has only bosonic fields as quantum vibrations, and that is the main reason that renders it to be unphysical.

The simplest way to incorporate fermions into the theory considers the presence of an extra symmetry at the level of the Polyakov action. A symmetry that relates a fermionic field for each bosonic field present in the two-dimensional string world-sheet. This is called supersymmetry. According with the Ramond-Neveu-Schwarz (R-NS) formalism, the bosonic fields $X^M(\sigma,\tau)$ are paired with fermionic partners $\psi^M(\sigma,\tau)$. In this case, the action of the bosonic string is modified by including the standard Dirac action. Therefore, the supersymmetric string action is,

$$S_{R-NS} = -\frac{1}{2\pi}\int d^2\sigma\left(\partial_\alpha X_M \partial^\alpha X^M + \psi^{-M}\rho^\alpha\partial_\alpha\psi_M\right),$$

where α is a world-sheet index and ρ^α are 2×2 matrices which obey the Dirac algebra. We shall use the representation given by,

$$\rho^0 \doteq \begin{bmatrix} 0 & -1 \\ 1 & 0 \end{bmatrix} \qquad \rho^1 \doteq \begin{bmatrix} 0 & 1 \\ 1 & 0 \end{bmatrix},$$

The two-dimensional spinor ψ^M has two Weyl components according to,

$$\psi^M = \begin{pmatrix} \psi_-^M \\ \psi_+^M \end{pmatrix}.$$

In the same way as the bosonic action, using the ligth-cone coordinates the fermionic part of the string action is rewriten as,

$$S_f = \frac{i}{\pi}\int d^2\sigma\left(\psi_-^M\partial_+\psi_{M-} + \psi_+^M\partial_-\psi_{M-}\right).$$

The whole action is invariant under supersymmetric transformations of the form,

$$\delta X^M = i\left(\epsilon_+\psi_-^M - \epsilon_-\psi^M\right),$$

$$\delta\psi_-^M = -2\partial_- X^M{}_{\epsilon+}$$

$$\delta\psi_+^M = 2\partial_+ X^M{}_{\epsilon-}$$

where $\epsilon_{\pm}$ are infinitesimal supersymmetric parameters corresponding to constant two-dimensional Majorana spinors.

Equations of Motion and Boundary Conditions

The equations of motion with respect to X^i and ψ^i (notice that there is again just (D – 2) degrees of freedom) are given respectively by the wave equation in $\partial_+\partial_- X^M = 0$ and the Dirac equations,

$$\partial_{\pm}\psi^M_{\mp} = 0.$$

The corresponding boundary terms for the fermionic part is given by,

$$\delta S = -T\int d\tau\left(\psi^M_+\delta\psi_{M+} - \psi^M_-\delta\psi_{M-}\right)\Big|_{\sigma=\pi} - \left(\psi^M_+\delta\psi_{M+} - \psi^M_-\delta\psi_{M-}\right)\Big|_{\sigma=0},$$

which vanishes in several different ways. Each way establishes the presence of different uncorrelated sectors of the theory as we shall see shortly.

Boundary Conditions for the Open String

Let us consider an open string, i.e., a string which endpoints are not coincident, and let us analyze the boundary conditions on the fermionic component of the action. Being an open string, both boundary terms in the variation of the action, must vanish separately. This is achieved if,

$$\psi^M_+ = \pm\psi^M_-,$$

which just reflects the very well known fact that fermions need a double rotation to recover their initial sign. Let us say that for two of the endpoints at $\sigma = 0$, both Weyl spinors are coincident, i.e., $\psi^M_+\big|_{\sigma=0} = \psi^M_-\big|_{\sigma=0}$. For the other endpoint there are two possible cases.

Periodic boundary conditions, $\psi^M_+\big|_{\sigma=\pi} = \psi^M_-\big|_{\sigma=\pi}$, are commonly called Ramond (R) boundary conditions. A solution satisfying this condition is said to belong to the R-sector, and it reads,

$$\psi^M_-(\sigma,\tau) = \frac{1}{\sqrt{2}}\sum_{n\in Z} d^M_n\, e^{-in\sigma^-},$$

$$\psi^M_+(\sigma,\tau) = \frac{1}{\sqrt{2}}\sum_{n\in Z} d^M_n\, e^{-in\sigma^+}.$$

The Majorana conditions requires these expansions to be real and hence $d^M_{-n} = d^{M\dagger}_n$. Notice that the index runs over integer numbers.

Anti-periodic boundary conditions, $\psi_{+}^{M}\big|_{\sigma=\pi} = -\psi_{-}^{M}\big|_{\sigma=\pi}$, are referred as Neveu-Schwarz (NS) boundary conditions. The corresponding solutions belong to the NS-sector, and are of the form,

$$\psi_{-}^{M}(\sigma,\tau) = \frac{1}{2}\sum_{r\in Z+\frac{1}{2}} d_r^M e^{-ir\sigma^-},$$

$$\psi_{+}^{M}(\sigma,\tau) = \frac{1}{2}\sum_{r\in Z+\frac{1}{2}} d_r^M e^{-ir\sigma^+}.$$

Therefore, the supersymmetric open string solutions are separated into two different sectors. Space-time bosons are constructed in the NS-sector, while space-time fermions are constructed in the R-sector.

Boundary Conditions for the Closed String

As in the bosonic string, the closed superstring solution can be represented in terms of left and right movers. On each of them is possible to impose periodicity or anti-periodicity boundary conditions, $\psi_{\pm}(\sigma) = \pm\psi_{\pm}(\sigma+\pi)$. This establishes four ways to determine a solution of the equations of motion. For instance, if both left and right movers satisfy NS boundary conditions, we say that the corresponding solution belongs to the Neveu-Schwartz-Neveu-Schwartz (NS-NS) sector. Hence, the four sectors are NS-NS, NS-R, R-NS and R-R. Bosonic fields arise from the NS-NS and R-R sector, while fermions belong to the R-NS and NS-R sectors. The solutions are denoted as,

$$\psi_{-}^{M}(\sigma,\tau) = \frac{1}{2}\sum_{n\in Z} d_n^M e^{-2in\sigma^-}, \psi_{-}^{M}(\sigma,\tau) = \frac{1}{2}\sum_{r\in Z+\frac{1}{2}} b_r^M e^{-2ir\sigma^-},$$

for the right movers with R and NS boundary conditions respectively, and,

$$\psi_{+}^{M}(\sigma,\tau) = \frac{1}{2}\sum_{n\in Z} \tilde{d}_n^M e^{-2in\sigma^-}, \psi_{+}^{M}(\sigma,\tau) = \frac{1}{2}\sum_{r\in Z+\frac{1}{2}} \tilde{b}_r^M e^{-2ir\sigma^-},$$

for the left movers. The specific fields arising as quantum modes of these solutions, will depend on which type of spinor we consider in the the supersymmetric action. There are two possibilities: chiral or non-chiral. So, there are at least two different superstring theories consisting on closed strings.

Canonical Quantization of the Supersymmetric String

As in the bosonic string case, we shall take a short-way and we will proceed to quantize the superstring by using the canonical quantization procedure, which means that Fourier modes are promoted to be operators satisfying some algebra. For the bosonic modes, the commutation relations are the same as in the bosonic string. For the fermionic modes we have that,

$$\left\{b_r^M, b_s^N\right\} = \eta^{MN}\delta_{r+s,0}$$

and similar expressions for the left movers. The reader should observe that such relations are precisely the algebra of raising and lowering operators in a harmonic oscillator satisfying the fermi statistics. Such modes act as creator and annihilator operators. By acting them on a suitable defined background state, we can describe the quantized string vibrations. Let us then define the background state $|\Omega\rangle_R$ for the right movers. For $n \geq 0$, $|\Omega\rangle_R$ is annihilated by the operators α_n^M and d_n^M in the R sector, while for the NS-sector, the annihilation comes from operator b_r^M, for r > 0. Similarly, for the left movers, we have that for the R sector, $\tilde{\alpha}_n^M|0\rangle_L = d_n^M|0\rangle_L = 0$, while for the NS-sector, $\tilde{b}_r^M|\Omega\rangle_L = 0$, for positive indices.

Therefore, excited states of the closed string are constructed by acting negative modes operators on the ground state $|\Omega\rangle_R \otimes |\Omega\rangle_L$. It turns out that the ground state for left or right movers is degenerated in the R-sector and unique in the NS-sector. In order to construct the respective excited states for the closed string, let us start by analyzing the construction on some left or right movers, which is equivalent to study the spectrum of the open string. In the NS sector, the ground state is unique, since all rising operators increment its mass. This follows, as in the bosonic case, from the energy momentum tensor. Classically, the energy-momentum tensor nonzero components read,

$$T_{++} = \partial_+ X_M \partial_+ X^M + \frac{1}{2}\psi_+^M \partial_+ \psi_{+M},$$

$$T_{--} = \partial_- X_M \partial_- X^M + \frac{1}{2}\psi_-^M \partial_- \psi_{-M},$$

which vanishes by the super Weyl symmetry. Hence, by substituting the solutions of the equations of motion, these energy-momentum tensor components can be written in terms of super Virasoro generators L_m and G_r satisfying some algebra. The generators L_m, with m an integer, are given by,

$$L_m = \frac{1}{2}\sum_m \alpha_{m-n}\cdot\alpha_m + \frac{1}{4}\sum_r (2r-m)\psi_{m-r}\cdot\psi_r + a\delta_{m,0},$$

where the product of modes operators are already ordered. The constant a is, as in the bosonic case, a measure of the zero point energy. For the R-sector, fermionic fields contribute with a same quantity than bosons, but with an opposite sign, i.e.,

$$a = a_b + a_f = -\frac{1}{12}(D-2)(1-1) = 0 \cdot$$

For the NS-sector, the situation is different. Being the modes labeled by fractional indices, the zero point energy contribution reads $a_f = -\frac{1}{48}(D-2)$, from which the total contribution to the zero point energy is −1/2.

Therefore, the mass is computed from the zero mode generator L_0, which in the NS-sector, is given by,

$$\alpha' M^2 = \sum_{n \in Z} \alpha_{-n} \cdot \alpha_n + \sum_{r \in Z+\frac{1}{2}} r b_{-r} \cdot b_r - \frac{1}{2},$$

A generic excited state is constructed by arbitrary values of N_B and N_F and that since all the oscillators transform as space-time vectors then all excited sates are bosons. In the R sector the ground state is degenerated since the operators d_0^M can act without changing the mass, which is given by,

$$\alpha' M^2 = \sum_{n \in Z} \alpha_{-n} \cdot \alpha_n + \sum_{n \in Z} n d_{-n} \cdot d_n.$$

This follows from the fact that zero modes d_0 satisfy the algebra $\{d_0^M, d_0^N\} = \eta^{MN}$ yielding the non-existence of a unique solution for $d_0^M |\Omega\rangle = 0$. Therefore, the set of ground states in the R sector must transform as a representation of Dirac algebra,

$$d_0^M |a\rangle = \frac{1}{\sqrt{2}} \Gamma_{ab}^M |b\rangle,$$

where Γ is a Dirac matrix. This implies that the R-sector ground state is actually a space-time fermion.

The Superstring Spectrum

For the NS sector we have seen that there is a negative-quadratic-mass state which is identified as the tachyon. Consistency requires the tachyon to be projected out. This can be achieved if we can project some states out, and the remaining states are self-contained, i.e., that by the action of any operator, the resulting state is contained among that set of states. It turns out that such projection exists and it is called the GSO projection (Gliozzi, Scherk and Olive). Basically the action of the GSO projector on the NS and R sector is to take out a half of the states arising at different energy values of the string.

Being more explicit, the action of the GSO projector on NS states is given by $(-1)^F$, with F being the world-sheet fermion number. Negative states under the GSO projector are removed. Then, the tachyonic state is projected out (the background state has fermion number F = 1) and the first state which survives GSO projection is $\psi_{1/2}^M |\Omega\rangle$, with M = 1, . . ., 8. The coordinates 0, 9 represent non-physical degrees of freedom. It is possible to show that this state is a massless gauge boson transforming in the adjoint of U(1) gauge group, and it is expressed as the vectorial representation 8_v of the little Lorentz group SO(8).

For the R sector, the energy of ground state is zero, implying there is a degeneration in the energy levels on states given by the zero mode ψ_0^M. In the light-cone gauge (or after considering the effect of ghosts) we have eight zero modes which can be used to construct 4 raising and 4 lowering operators by a linear combination of two zero modes. Applying the raising operator to a bosonic vacuum $|\Omega\rangle$ we generate $2^4 = 16$ states. Now, on the one hand, there are at least 16 supersymmetry generators, corresponding to the Weyl components of the full spinor, which has 32 degrees of freedom. This spinor is a Majorana-Weyl spinor, which splits it into two real spinors with opposite chirality, i.e., $32 = 16 \oplus 16'$. Then, under spinorial representations of the Lorentz group SO(1, 9), we have the following decomposition,

$$SO(1,\ 9) \rightarrow SO(1,\ 1) \times SO(8)$$

$$16 \rightarrow \left(\frac{1}{2},\ 8_s\right) \oplus \left(-\frac{1}{2}, 8_c\right).$$

On the other hand, there is one constraint (Dirac equation) that threw out representations with quantum number $-\frac{1}{2}$ of SO(1, 1). That is why we kept just 16 states that correspond to those constructed by raising operators described above.

Table: Open superstring spectrum.

Sector	State	Level	M^2/α	SO(8) rep.	Field	GSO
NS	$\|\Omega\rangle_{NS}$	$N_b = N_f = 0$	-1/2	-	Tachyon	-
NS	$\psi_{-1}^M \|\Omega\rangle_{NS}$	$N_b = 0,\ N_f = 1$	0	8_V	Gauge boson	+
R	$\|\Omega\rangle_R$	$N_b = N_f = 0$	0	8_C	Gauginos	-
R	$\|\Omega\rangle_R$	$N_b = N_f = 0$	0	8_S	Gauginos	+

Therefore, GSO projection in RR sector is basically the application of the chiral matrix $\bar{\Gamma} = i\Gamma^0\Gamma^1 \ldots \Gamma^9$ to the spinor states. Because the states in the 16 of SO(8) are spinors, the action of $\bar{\Gamma}$ split up them into two Weyl fermions, which are representations of SO(8) given by 8_s and 8_c with different chirality. After GSO projection, the R-sector contains 16 spinorial states in the 8s of SO(8) that represents the super-partners of 8_v in the NS sector (gauginos). The spectrum of the open string is shown in open superstring spectrum table.

Closed Superstring Spectrum

Once the spectrum of the open string has been computed, it is straightforward to compute the closed string spectrum. The resulting fields are constructed by the tensorial product of the corresponding states for the left and right movers. However, there are two different ways to do it. The theory is called type II A if left and right-moving R-sector ground states are chosen to have the opposite chirality (which reflects in the fact that the two R-sectors have opposite parity. In the type II B left and right-moving sector ground states have the same chirality. These two consistent ways to construct a superstring theory spectrum, are summarized in tables below.

Table: Type II A superstring spectrum.

Sector	State	SO(8) rep.	Fields
NS-NS	$\psi^{M}_{-1}\tilde{\psi}^{N}_{-1}\|\Omega\rangle_L \otimes \|\Omega\rangle_R$	$8_v \otimes 8_v = 1 \oplus 26 \oplus 35$	G_{MN}, B_{MN}, ϕ
R-R	$\|+\rangle_L \otimes \|+\rangle_R$	$8_s \otimes 8_c = 8_v \oplus 56_t 8_v$	$C_{(1)}, C_{(3)}$
NS-R	$\psi^{M}_{-1}\|\Omega\rangle_L \otimes \|-\rangle$	$8_v \otimes 8_c = 8_s \oplus 56_c$	$\lambda^{\alpha}, \psi^{\alpha}_{M}$
R-NS	$\|+\rangle_L \otimes \tilde{\psi}^{M}_{-1}\|\Omega\rangle_R$	$8_s \otimes 8_v = 8_c \oplus 56_s$	$\lambda^{\dot{\alpha}}, \psi^{\dot{\alpha}}_{M}$

Table: Type II B superstring spectrum.

Sector	State	SO(8) rep.	Fields
NS-NS	$\psi^{M}_{-1}\tilde{\psi}^{N}_{-1}\|\Omega\rangle_L \otimes \|\Omega\rangle_R$	$8_v \otimes 8_v = 1 \oplus 26 \oplus 35$	G_{MN}, B_{MN}, ϕ
R-R	$\|+\rangle_L \otimes \|+\rangle_R$	$8_s \otimes 8_s = 1 \oplus 26 \oplus 35$	$C_{(0)}, C_{(2)}$
NS-R	$\psi^{M}_{-1}\|\Omega\rangle_L \otimes \|+\rangle$	$8_v \otimes 8_s = 8_s \oplus 56_s$	$\lambda^{\alpha}, \psi^{\alpha}_{M}$
R-NS	$\|+\rangle_L \otimes \tilde{\psi}^{M}_{-1}\|\Omega\rangle_R$	Same as NS-R	

The Type II A Superstring Spectrum

The representation of the complete states is obtained by tensoring the representation of the left and right movers. The NS-NS sector contains a symmetric tensor of two index corresponding to a graviton (G_{MN}), an antisymmetric tensor or 2-form (B_{MN}) and a dilaton (Φ). The R-R sector contains a set of completely antisymmetric tensors (p-forms). In particular a 1-form (C_M) and a 3-form (C_3). It is sometimes convenient to introduce the Hodge duals of these, which are a 5-form (C_5) and a 7-form (C_7). Finally, it is also useful to introduce a 9-form C_9, with no dynamics. The space-time fermions arise from the R-NS and NS-R sectors which contain the gravitino (56 states) and a dilatino (8 states) with opposite chirality. The R-NS and NS-R sectors contain fermions, which are a spin 3/2 gravitino and a spin 1/2 dilatino, each of these fermions with opposite chirality.

The Type II B Superstring Spectrum

For this case, we choose left and right movers in R-sector to have the same chirality (in the 8_s and 8_s of SO(8)). The NS-NS sector contains the same fields as in the Type II A case. The R-NS and NS-R sectors contain fermions with the same chirality. Basically they are two copies of fermions in the NS-R sector in type II A. The RR sector contains a 0-form C_0 (scalar), a 2-form C_2 and a 4-form C_4^{+} with a self-dual field strength. It is sometimes convenient to introduce the Hodge duals of these, which are a 6-form B_6, and 8-form C_8. Finally, it is also useful to introduce a 10-form C_{10}, which does not have any propagating degrees of freedom, since it has no space-time kinetic term. The Type II A and II B superstring theories have as low energy description (when just massless modes are considered) the Types II A and II B supergravity (SUGRA) versions. In other words, II refers to supersymmetry $\mathcal{N} = 2$ and we have two gravitinos in each supergravity multiplet with D = 10[14].

Effective Field Theory

Effective field theories are just general quantum field theories based on the principles of relativity and quantum mechanics but without the restriction of renormalizability. Any term allowed by the assumed symmetries of the theory is included in the action. The characteristic feature of an effective field theory is that it can be valid only up to a certain energy scale Λ, which must be above the energy scale directly accessible by experiment for the theory to be useful. In contrast, fundamental theories are candidates for theories that describe Nature at any scale. We will never know whether a given candidate for a fundamental theory is the theory that describes Nature at all energies. Whether such a theory is adopted as the 'Standard Model' then depends on aesthetic criteria such as naturalness and theoretical simplicity. Effective theories are more adapted to describe what we can know.

A perturbative power-counting analysis shows that fundamental theories must not contain operators with mass dimension larger than four in their Lagrangian, while effective theories contain all possible higher-dimensional operators. Since the effect of short-distance physics looks local to an observer performing experiments at energies $E \ll \Lambda$, the ignorance of this physics can be parametrized by the values of the coupling constants of all possible local operators. Effective theories are predictive, since an operator of dimension d contributes at most of order $(E/\Lambda)^{d-4} \ll 1$ to a dimensionless observable, hence only a finite number of unknown couplings are relevant to achieve a set precision. While so-called non-renormalizable interactions (d > 4) thus appear naturally in effective field theories, super-renormalizable interactions (d < 4, such as mass terms) are problematic, since their couplings are generically large, of order Λ^{4-d}. Mass terms of order Λ^2, however, describe particles that cannot be part of the effective theory. Thus super-renormalizable interactions should be protected by symmetries (chiral, gauge, etc.) to avoid a hierarchy problem.

Integrating Out a Heavy Quark

Suppose there is nothing else in the world than gluons and six flavours of quarks one of which, the top quark, is much heavier than the others. The fundamental theory is,

$$\mathcal{L}=-\frac{1}{4}G^{A}_{\mu\nu}G^{A\mu\nu}+\sum_{f=1}^{5}\bar{\psi}_f(i\not{D}-m_f)\psi_f+\bar{Q}(i\not{D}-m_f)Q.$$

At energies $E \ll m_Q$ much below the top quark mass, physics is described by an effective field theory without the top quark field Q but containing all higher-dimensional operators built from ψ_f and the gluon field constrained only by the SU(3) gauge symmetry. Since the fundamental theory is known and weakly coupled at the scale m_Q, the effective field theory Lagrangian can be calculated order by order by a procedure called 'matching'. The simplest diagram containing the top quark is the gluon vacuum polarization at one loop. Expand the result in external momentum $q^2 \ll m_Q^2$. The result can be reproduced in an effective theory without the top quark with Lagrangian,

$$\mathcal{L}=-\frac{1}{4}\left[1-\frac{\alpha_s T_F}{3\pi}\ln\frac{m_Q^2}{\mu^2}\right]G^{A}_{\mu\nu}G^{A\mu\nu}+\sum_{f=1}^{5}\bar{\psi}_f(i\not{D}-m_f)\psi_f+\frac{\alpha_s T_f}{60\pi m_Q^2}G^{A}_{\mu\nu}D^2G^{A\mu\nu}+\ldots$$

This simple example illustrates a few important points:

- The couplings of the dimension 4 terms have been modified, here the gluon kinetic energy term, and a series of higher-dimensional operators has been generated. Matching the gluon three-point function would generate another dimension 6 operator $f^{ABC}G^{A}_{\mu\nu}G^{B\mu\alpha}G^{C\nu}_{\alpha}$, and so on.
- To express the low-energy Lagrangian in terms of canonically normalized fields, we must rescale the gluon field:

$$\hat{A}^{A}_{\mu}=\left(1-\frac{\alpha_s T_f}{6\pi}\ln\left(m_Q^2/\mu^2\right)A^{A}_{\mu}\right).$$

- Then the effective quark–gluon coupling is,

$$g_s\bar{\psi}_f\not{A}\psi_f=\frac{g_s}{1-\frac{\alpha_s T_f}{6\pi}\ln\frac{m_Q^2}{\mu^2}}\bar{\psi}_f\hat{\not{A}}\psi_f\equiv\hat{g}_s\bar{\psi}_f\hat{\not{A}}\psi_f.$$

- The strong coupling in the effective theory is not the same as in the fundamental theory. However, $\alpha_s(m_Q)=\hat{\alpha}_s(m_Q)$ in the present approximation.

- If $\mu^2 \frac{d\alpha_s}{d\mu^2} = -\beta_0^{(6)} \alpha_s^2 / (4\pi)$ with $\beta_0^{(6)} = 11 - \frac{2}{3} \cdot 6$, then $\mu^2 \frac{d\hat{\alpha}_s}{d\mu^2} = -\beta_0^{(5)} \hat{\alpha}_s^2 / (4\pi)$ with $\beta_0^{(5)} = 11 - \frac{2}{3} \cdot 5$ just as expected. At scales $\mu \ll m_Q$ one must use the effective five-flavour Lagrangian to sum logarithms $\left[\alpha_s \ln\left(m_Q^2 / \mu^2\right)\right]^n$ to all orders in perturbation theory, which otherwise ruin perturbation theory even when α_s is small.

If we did not know the fundamental theory, we could not tell the existence of the top quark from the dimension 4 terms, since their effect is hidden in the value of the low-energy strong coupling $\hat{\alpha}_s$ that is determined experimentally. However, as experiments approach the top quark scale, neglecting the dimension 6 operator becomes a poorer and poorer approximation, and from the failure to describe experimental data accurately we could infer the existence of a scale $\Lambda \approx m_Q$, where the effective theory breaks down. We might imagine measuring the couplings of the dimension 6 operators and from their value derive the hypothesis that there must be a yet undiscovered heavy quark species.

Effective Weak Interactions

The weak interactions were described by a set of dimension 6 four-fermion operators. The non-renormalizability of these interactions provided strong motivation for searching for a fundamental theory of the weak interactions which resulted in the discovery of the SU(2)×U(1) gauge structure now included in the Standard Model. The value of the Fermi constant GF together with naturalness arguments points towards the 100 GeV scale as the limiting scale of Fermi's theory of the weak interaction; indeed the W and Z bosons have masses of this order.

Now that the gauge sector of the Standard Model is established, we use the reverse procedure. Since all flavour-changing processes observed up to now occur in processes where the external scales are at or below the bottom quark mass $m_b \ll M_W, M_Z, m_t$, we can use an effective field theory approximation to the Standard Model where the W and Z boson, and the top quark have been integrated out. The result is five-flavour QCD and an extended version of Fermi's four-fermion interactions for the weak force:

$$\mathcal{L}_{SM} \rightarrow \mathcal{L}_{QCD_5 + QED} + \mathcal{L}_{eff,weak}.$$

The leading term in the effective weak interactions $\mathcal{L}_{eff,weak}$ is a dimension 6 interaction, which explains why the weak interactions are weak at low energies $E \ll M_W$, the effective coupling is $g_\omega^2 (E / M_W)^2$ but stronger than the electromagnetic interactions at the Fermi scale 100 GeV.

Supersymmetry

Supersymmetry in particle physics a symmetry between fermions (subatomic particles with half-integer values of intrinsic angular momentum, or spin) and bosons (particles with integer values of spin). Supersymmetry is a complex mathematical framework based on the theory of group transformations that was developed beginning in the early 1970s to understand at a more-fundamental level the burgeoning number of subatomic particles being produced in high-energy particle accelerator experiments. It has evolved to address internal inconsistencies that arose in attempts to unify the forces in the Standard Model of particle physics. Supersymmetry is an essential feature of supergravity, the quantum field theory of the gravitational force, and of string theory, an ambitious attempt to provide a self-consistent quantum theory unifying all particles and forces in nature.

Supersymmetric models often are easier to solve than non-supersymmetric ones since they are more constrained by the higher degree of symmetry. Thus they may serve as toy models where certain analytic results can be obtained and may serve as a qualitative guide to the behaviour of more realistic theories. For example the study of supersymmetric versions of QCD have given quite some insights in the strong coupling dynamics responsible for phenomena like quark confinement. In this type of studies the basic property is a duality (a mapping) between a weakly and a strongly coupled theory. It seems that dualities are difficult to realise in non-supersymmetric theories but are rather easily present in supersymmetric ones. Supersymmetry has also appeared outside the realm of elementary particle physics and has found applications in condensed matter systems, in particular in the study of disordered systems.

Spinors and the Poincaré Group

We begin with a review of the Lorentz and Poincaré groups and spinors in four-dimensional Minkowski space. The signature is taken to be +,-,-,- so that $p^2 = +m^2$ and μ, ν, . . . always are space-time indices, while i, j, . . . are only space indices. Then the metric $g_{\mu\nu}$ is diagonal with $g_{00} = 1$, $g_{11} = g_{22} = g_{33} = -1$.

The Lorentz and Poincaré Groups

The Lorentz group has six generators, three rotations J_i and three boosts K_i , i = 1, 2, 3 with commutation relations,

$$\left[J_i,J_j\right]=i\in_{ijk} J_k,\left[K_i,K_j\right]=-i\in_{ijk} J_k,\left[J_i,K_j\right]=i\in_{ijk} K_j$$

To identify the mathematical structure and to construct representations of this algebra one introduces the linear combinations,

$$J_j^{\pm}=\frac{1}{2}\left(J_j \pm iK_j\right)$$

in terms of which the algebra separates into two commuting SU(2) algebras:

$$\left[J_i^{\pm}, J_j^{\pm}\right] = i \in_{ijk} J_k^{\pm}, \left[J_i^{\pm}, J_j^{\mp}\right] = 0.$$

These generators are not hermitian however, and we see that the Lorentz group is a complexified version of SU(2) × SU(2): this group is Sl(2, C) . (More precisely, Sl(2, C) is the universal cover of the Lorentz group, just as SU(2) is the universal cover of SO(3).) To see that this group is really Sl(2, C) is easy: introduce the four 2 × 2 matrices σ_μ where σ_0 is the identity matrix and σ_i , i = 1, 2, 3 are the three Pauli matrices. (We always write the Pauli matrices with a lower index i, while $\sigma^0 = \sigma_0$ and $\sigma^i = -\sigma_i$.) Then for every four-vector x^μ the 2 × 2 matrix $x^\mu \sigma_\mu$ is hermitian and has determinant equal to $x^\mu x_\mu$ which is a Lorentz invariant. Hence a Lorentz transformation preserves the determinant and the hermiticity of this matrix, and thus must act as $x^\mu \sigma_\mu \to A x^\mu \sigma_\mu A^\dagger$ with $|\det A| = 1$. We see that up to an irrelevant phase, A is a complex 2 × 2 matrix of unit determinant, i.e. an element of Sl(2, C) . This establishes the mapping between an element of the Lorentz group and the group Sl(2, C) .

The Poincaré group contains, in addition to the Lorentz transformations, also the translations. More precisely it is a semi-direct product of the Lorentz-group and the group of translations in space-time. The generators of the translations are usually denoted P_μ. In addition to the commutators of the Lorentz generators J_i (rotations) and K_i (boosts) one has the following commutation relations involving the P_μ:

$$\left[P_\mu, P_\nu\right] = 0$$
$$\left[J_j, P_j\right] = i \in_{ijk} P_k, \left[J_i, P_0\right] = 0, \left[K_i, P_j\right] = -iP_0, \left[K_i, P_0\right] = -iP_j,$$

which state that translations commute among themselves, that the P_i are a vector and P_0 a scalar under space rotations and how P_i and P_0 mix under a boost. The Lorentz and Poincaré algebras are often written in a more covariant looking, but less intuitive form. One defines the Lorentz generators $M_{\mu\nu} = -M_{\nu\mu}$ as $M_{0i} = K_i$ and $M_{ij} = \in_{ijk} J_k$. Then the full Poincaré algebra reads,

$$\left[P_\mu, P_\nu\right] = 0,$$
$$\left[M_{\mu\nu}, M_{\rho\sigma}\right] = ig_{\nu\rho} M_{\mu\sigma} - ig_{\mu\rho} M_{\nu\sigma} - ig_{\nu\sigma} M_{\mu\rho} + ig_{\mu\sigma} M_{\nu\rho},$$
$$\left[M_{\mu\nu}, P_\rho\right] = -ig_{\rho\mu} P_\nu + ig_{\rho\nu} P_\mu.$$

Spinors

Two-Component Spinors

There are various equivalent ways to introduce spinors. Here we define spinors as the objects carrying the basic representation of Sl(2, C) . Since elements of Sl(2, C) are complex 2×2 matrices, a spinor is a two complex component object $\psi = \begin{pmatrix} \psi_1 \\ \psi_2 \end{pmatrix}$ transforming under an element $\mathcal{M} = \begin{pmatrix} \alpha & \beta \\ \gamma & \delta \end{pmatrix} \in \mathrm{Sl}(2, C)$ as,

$$\psi_\alpha \to \psi'_\alpha = \mathcal{M}_\alpha^\beta \, \psi_\beta,$$

with α, β = 1, 2 labeling the components. Now, unlike for SU(2), for Sl(2, C) a representation and its complex conjugate are not equivalent. $\mathcal{M}$ and $\mathcal{M}^*$ give inequivalent representations. A two-component object $\bar{\psi}$ transforming as,

$$\psi_\alpha \to \psi'_\alpha = \mathcal{M}_\alpha^\beta \, \psi_\beta,$$

is called a dotted spinor, while the above ψ is called an undotted one. Comparing the complex conjugate of $\psi_\alpha \to \psi'_\alpha = \mathcal{M}_\alpha^\beta \, \psi_\beta$, with $\psi_\alpha \to \psi'_\alpha = \mathcal{M}_\alpha^\beta \, \psi_\beta$, we see that we can identify $\bar{\psi}_{\dot{\alpha}}$ with $(\psi_\alpha)^*$. The representation carried by the ψ_α is called $(\frac{1}{2}, 0)$ (matrices $\mathcal{M}$) and the one carried by the $\bar{\psi}_{\dot{\alpha}}$ is called $(0, \frac{1}{2})$ (matrices $\mathcal{M}^*$). They are both irreducible. Now, any Sl(2, C) matrix can be written as,

$$\mathcal{M} = \exp\left(a_j \sigma_j + i b_j \sigma_j\right)$$
$$\mathcal{M}^* = \exp\left(a_j \sigma_j^* - i b_j \sigma_j^*\right).$$

This explicitly displays the generators as the spin $\frac{1}{2}$ representation of the complexified SU(2), in accordance with $J_j^\pm = \frac{1}{2}\left(J_j \pm i K_j\right)$. It proves very useful to now introduce some notations and conventions. We intoduce the antisymmetric two-index tensors $\epsilon^{\alpha\beta}$ and $\epsilon_{\alpha\beta}$,

$$\epsilon^{\alpha\beta} = \epsilon^{\dot{\alpha}\dot{\beta}} = \begin{pmatrix} 0 & 1 \\ -1 & 0 \end{pmatrix}, \epsilon_{\alpha\beta} = \epsilon_{\dot{\alpha}\dot{\beta}} = \begin{pmatrix} 0 & -1 \\ 1 & 0 \end{pmatrix},$$

which are used to raise and lower indices as follows:

$$\psi^\alpha = \epsilon^{\alpha\beta} \psi^\beta, \psi_\alpha = \epsilon_{\alpha\beta} \psi^\beta, \bar{\psi}^{\dot{\alpha}} = \epsilon^{\dot{\alpha}\dot{\beta}} \bar{\psi}_{\dot{\beta}}, \bar{\psi}_{\dot{\alpha}} = \epsilon_{\dot{\alpha}\dot{\beta}} \bar{\psi}^{\dot{\beta}}.$$

One can then easily show that the transformation under an element $\mathcal{M}$ of Sl(2, C) is:

$$\psi'^\alpha = \psi^\beta \left(\mathcal{M}^{-1}\right)_\beta^\alpha \text{ and } \bar{\psi}'^{\dot{\alpha}} = \bar{\psi}^{\dot{\beta}} \left(\mathcal{M}^{*-1}\right)_{\dot{\beta}}^{\dot{\alpha}}.$$

The four σ_μ matrices introduced above naturally have a dotted and an undotted index. Recalling that our signature is +,-,-,- we have,

$$\left(\sigma_\mu\right)_{\alpha\dot{\alpha}} = \left(1, -\sigma_i\right)_{\alpha\dot{\alpha}}.$$

Raising the indices using the $\in$ tensors yields,

$$\left(\bar{\sigma}^\mu\right)^{\dot{\alpha}\alpha} = \in^{\dot{\alpha}\dot{\beta}} \in^{\alpha\beta} \left(\sigma^\mu\right)_{\beta\dot{\beta}} = \left(1, +\sigma_i\right)^{\dot{\alpha}\alpha}.$$

Whenever we consider expressions involving more than one spinor we have to remember that spinors anticommute. Hence (with two-component spinors) $\psi_1\chi_2 = -\chi_2\psi_1$,, as well as $\psi_1\bar{\chi}_{\dot{2}} = -\chi_{\dot{2}}\psi_1$ etc. The scalar products $\psi\chi$ and $\bar{\psi}\bar{\chi}$ are defined as,

$$\psi\chi \equiv \psi^\alpha\chi_\alpha = \in^{\alpha\beta}\psi_\beta\chi_\alpha = -\in^{\alpha\beta}\psi_\alpha\chi_\beta = -\psi_\alpha\chi^\alpha = \chi^\alpha\psi_\alpha = \chi\psi$$

$$\bar{\psi}\bar{\chi} \equiv \bar{\psi}_{\dot{\alpha}}\bar{\chi}^{\dot{\alpha}} = \ldots = \bar{\chi}_{\dot{\alpha}}\bar{\psi}^{\dot{\alpha}}$$

$$\left(\psi\chi\right)^\dagger = \bar{\chi}_{\dot{\alpha}}\bar{\psi}^{\dot{\alpha}} = \bar{\chi}\bar{\psi} = \bar{\psi}\bar{\chi}.$$

By convention undotted indices are always contracted from upper left to lower right, while dotted indices are always contracted from lower left to upper right. Note however that this rule does not apply when rising or lowering spinor indices with the $\in$-tensor. With this rule we also have,

$$\psi\sigma^\mu\bar{\chi} = \psi^\alpha\sigma^\mu_{\alpha\dot{\beta}}\bar{\chi}^{\dot{\beta}},\ \bar{\psi}\bar{\sigma}^\mu\chi = \bar{\psi}_{\dot{\alpha}}\bar{\sigma}^{\mu\dot{\alpha}\beta}\chi_\beta.$$

One can then prove a certain amount of useful identities which we summarise here,

$$\chi\sigma^\mu\bar{\psi} = -\bar{\psi}\bar{\sigma}^\mu\chi,\ \chi\sigma^\mu\bar{\sigma}^\nu\psi = \psi\sigma^\nu\bar{\sigma}^\mu\chi$$

$$\left(\chi\sigma^\mu\bar{\psi}\right)^\dagger = \psi\sigma^\mu\bar{\chi},\ \left(\chi\sigma^\mu\bar{\sigma}^\nu\psi\right)^\dagger = \bar{\psi}\bar{\sigma}^\nu\sigma^\mu\bar{\chi}$$

$$\psi\chi = \chi\psi,\ \bar{\psi}\bar{\chi} = \bar{\chi}\bar{\psi},\ \left(\psi\chi\right)^\dagger = \bar{\psi}\bar{\chi}.$$

Dirac Spinors

One introduces the Dirac matrices in the Weyl representation as,

$$\gamma^\mu = \begin{pmatrix} 0 & \sigma^\mu \\ \bar{\sigma}^\mu & 0 \end{pmatrix},\ \gamma_5 = i\gamma^0\gamma^1\gamma^2\gamma^3 = \begin{pmatrix} 1 & 0 \\ 0 & -1 \end{pmatrix}$$

A four-component Dirac spinor is made from a two-component undotted and a two-component dotted spinor as $\begin{pmatrix} \psi_\alpha \\ \bar{\chi}^{\dot{\alpha}} \end{pmatrix}$. Clearly it transforms as the reducible $\left(\frac{1}{2}, 0\right) \oplus \left(0, \frac{1}{2}\right)$ representation of the Lorentz group. Then $\begin{pmatrix} \psi_\alpha \\ 0 \end{pmatrix}$ and $\begin{pmatrix} 0 \\ \bar{\chi}^{\dot{\alpha}} \end{pmatrix}$ are chiral Dirac (or Weyl) spinors. A Majorana spinor is a Dirac spinor with $\chi \equiv \psi$, i.e. it is of the form $\begin{pmatrix} \psi_\alpha \\ \bar{\psi}^{\dot{\alpha}} \end{pmatrix}$. The Lorentz generators are,

$$\sum^{\mu\nu} = \frac{i}{2}\gamma^{\mu\nu},\ \gamma^{\mu\nu} = \frac{1}{2}\left(\gamma^\mu\gamma^\nu - \gamma^\nu\gamma^\mu\right) = \frac{1}{2}\begin{pmatrix} \sigma^\mu\bar{\sigma}^\nu - \sigma^\nu\bar{\sigma}^\mu & 0 \\ 0 & \bar{\sigma}^\mu\sigma^\nu - \bar{\sigma}^\nu\sigma^\mu \end{pmatrix}.$$

We see that indeed the undotted and dotted spinors transform separately, the generators being $i\sigma^{\mu\nu}$ for ψ_α and $i\bar{\sigma}^{\mu\nu}$ for $\bar{\psi}^{\dot{\alpha}}$ with,

$$\left(\sigma^{\mu\nu}\right)_\alpha^{\ \beta} = \frac{1}{4}\left(\sigma^\mu_{\alpha\dot{\gamma}}\bar{\sigma}^{\nu\dot{\gamma}\beta} - (\mu \leftrightarrow \nu)\right)$$

$$\left(\bar{\sigma}^{\mu\nu}\right)^{\dot{\alpha}}_{\ \dot{\beta}} = \frac{1}{4}\left(\bar{\sigma}^{\mu\dot{\alpha}\gamma}\sigma^\nu_{\gamma\dot{\beta}} - (\mu \leftrightarrow \nu)\right).$$

Casimirs: Mass and Helicity

A useful quantity is the Pauli-Lubanski vector,

$$W^\mu = \frac{1}{2} \in^{\mu\nu\rho\sigma} P_\nu M_{\rho\sigma},$$

which can be easily shown to commute with the P_μ and behaves as a four-vector under commutation with the Lorentz generators. It follows that $W^2 \equiv W^\mu W_\mu$ as well as $P^2 \equiv P_\mu P^\mu$ commute with all the generators, i.e. they are two (and the only two) Casimirs of the Poincaré group. For a massive particle one can go to the rest frame where $P_\mu = (m,0,0,0)$ and then $P^2 = m^2$ and $W^2 = -m^2 s(s+1)$ where s is the spin of the particle. The different states of this irreducible representation are distinguished by the value of p_i and of $M_{12} \equiv J_3 = S_3$ in the rest frame because in the rest frame only the spin and not the orbital part contributes to the angular momentum. In the above representation of dotted or undotted spinors one has of course $s = \frac{1}{2}$. For a massless particle $P^2 = 0$ and also $W^2 = 0$. We may take $P_\mu = (E,0,0,E)$ so that $W^\mu = M_{12}P^\mu$ with $M_{12} = \pm s$ being the helicity. For such massless particles s is fixed and the different states of this irreducible representation are distinguished by the sign of the helicity and by the values of p_i.

The Susy Algebra and its Representations

The Supersymmetry Algebra

We want to enlarge the Poincaré algebra by generators that transform either as undotted spinors or as dotted spinors $\bar{Q}^I_{\dot{\alpha}}$ under the Lorentz group and that commute with the translations. The extra index I = 1, . . .N labels the different spinorial generators in case there are more than one pair. This means that according to $\left(\sigma^{\mu\nu}\right)_{\alpha}^{\beta}=\frac{1}{4}\left(\sigma_{\alpha\dot{\gamma}}^{\mu}\bar{\sigma}^{\nu\dot{\gamma}\beta}-(\mu\leftrightarrow\nu)\right)$, $\left(\bar{\sigma}^{\mu\nu}\right)_{\dot{\beta}}^{\dot{\alpha}}=\frac{1}{4}\left(\bar{\sigma}^{\mu\dot{\alpha}\gamma}\sigma_{\gamma\dot{\beta}}^{\nu}-(\mu\leftrightarrow\nu)\right)$,

$$\begin{aligned}
&\left[P_{\mu},Q_{\alpha}^{I}\right]=0,\\
&\left[P_{\mu},\bar{Q}_{\dot{\alpha}}^{I}\right]=0,\\
&\left[M_{\mu\nu},Q_{\alpha}^{I}\right]=i\left(\sigma_{\mu\nu}\right)_{\alpha}^{\beta}Q_{\beta}^{I},\\
&\left[M_{\mu\nu},\bar{Q}^{I\dot{\alpha}}\right]=i\left(\bar{\sigma}_{\mu\nu}\right)_{\dot{\beta}}^{\dot{\alpha}}\bar{Q}^{I\dot{\beta}}.
\end{aligned}$$

In particular, $M_{12}\equiv J_3$ and thus $\left[J_3,Q_1^I\right]=\frac{1}{2}Q_1^I$ and $\left[J_3,Q_2^I\right]=-\frac{1}{2}Q_2^I$. Since $\bar{Q}^{I1}=-\left(Q_2^I\right)^\dagger$ and $\bar{Q}^{I2}=-\left(Q_1^I\right)^\dagger$ one similarly has $\left[J_3,\left(Q_2^I\right)^\dagger\right]=\frac{1}{2}\left(Q_2^I\right)^\dagger$ and $\left[J_3,\left(Q_1^I\right)^\dagger\right]=-\frac{1}{2}\left(Q_1^I\right)^\dagger$. We conclude that Q_1^I and $\left(Q_2^I\right)^\dagger$ rise the z-component of the spin (helicity) by half a unit, while Q_2^I and $\left(Q_1^I\right)^\dagger$ lower it by half a unit.

Since the Q_α^I transform in the $\left(\frac{1}{2},0\right)$ representation and the $\bar{Q}_{\dot{\alpha}}^I$ in the $\left(0,\frac{1}{2}\right)$, the anti-commutator of Q_α^I and $\bar{Q}_{\dot{\beta}}^J$ must transform as $\left(\frac{1}{2},\frac{1}{2}\right)$, i.e. as a four vector. The obvious candidate is Pμ so that we arrive at, $\left\{Q_\alpha^I,\bar{Q}_{\dot{\beta}}^J\right\}=2\sigma_{\alpha\dot{\beta}}^{\mu}P_\mu\delta^{IJ}$.

The δ^{IJ} can always be achieved by diagonalising an a priori arbitrary symmetric matrix and by rescaling the Q and $\bar{Q}$. Furthermore, since $\bar{Q}$ is the adjoint of Q, positivity of the Hilbert space excludes zero eigenvalues of this matrix. Finally,

$$\left\{Q_\alpha^I,Q_\beta^J\right\}=\epsilon_{\alpha\beta}Z^{IJ},\left\{\bar{Q}_{\dot{\alpha}}^I,\bar{Q}_{\dot{\beta}}^J\right\}=\epsilon_{\dot{\alpha}\dot{\beta}}\left(Z^{IJ}\right)^*.$$

The $Z^{IJ}=-Z^{JI}$ are central charges which means they commute with all generators of the full algebra. The simplest algebra has N = 1, i.e. there are no indices I, J and there is no possibility of central charges. This is the unextended susy algebra. If N > 1 one talks about extended supersymmetry. In the simplest extended case, N = 2, there is just one central charge $Z\equiv Z^{12}$.. From the algebraic point of view there is no limit on N, but we will see that with increasing N the theories also must contain particles of increasing spin and there seem to be no consistent quantum field theories with spins larger than one (without gravity) or larger than two (with gravity) leading to $N\le 4$, resp. $N\le 8$.

Some Basic Properties

Using the above susy algebra it is easy to establish some basic properties of supersymmetric theories. Since the full susy algebra contains the Poincaré algebra as a subalgebra, any representation of the full susy algebra also gives a representation of the Poincaré algebra, although in general a reducible one. Since each irreducible representation (of the type considered above) of the Poincaré algebra corresponds to a particle, an irreducible representation of the susy algebra in general corresponds to several particles. The corresponding states are related to each other by the Q^I_α and $\bar{Q}^J_{\dot\beta}$ and thus have spins differing by units of one half. They form a supermultiplet. By abuse of language we will call an irreducioble representation of the susy algebra simply a supermultiplet. Clearly, using the spin-statistics theorem, the Q and $\bar{Q}$ change bosons into fermions and vice versa. One then has:

All particles belonging to an irreducible representation of susy, i.e. within one supermultiplet, have the same mass. This is obvious since P^2 commutes with all generators of the susy algebra, i.e. it is still a Casimir operator. In a supersymmetric theory the energy P_0 is always positive. To see this, let $|\Phi\rangle$ be any state. Then by the positivity of the Hilbert space we have,

$$0 \le \left\| Q^I_\alpha |\Phi\rangle \right\|^2 + \left\| \left(Q^I_\alpha\right)^\dagger |\Phi\rangle \right\|^2 = \langle \Phi | \left(\left(Q^I_\alpha\right)^\dagger Q^I_\alpha + Q^I_\alpha \left(Q^I_\alpha\right)^\dagger |\Phi\rangle \right)$$

$$= \langle \Phi | \left\{ Q^I_\alpha, \bar{Q}^I_{\dot\alpha} \right\} | \Phi \rangle = 2\sigma^\mu_{\alpha\dot\alpha} \langle \Phi | P_\mu | \Phi \rangle$$

since $\left(Q^I_\alpha\right)^\dagger \equiv \bar{Q}^I_{\dot\alpha}$. Summing this over $\alpha \equiv \dot\alpha = 1,2$ and using tr $\sigma^\mu = 2\delta^{\mu 0}$ yields $0 \le 4\langle \Phi | P_0 | \Phi \rangle$, which was to be shown.

A supermultiplet always contains an equal number of bosonic and fermionic degrees of freedom. By degrees of freedom one means physical (positive norm) states. Hence a photon has two degrees of freedom corresponding to the two helicities +1 and −1 (the two polarizations). Let the fermion number be N_F equal one on a fermionic state and 0 on a bosonic one. Equivalently $(-)^{N_F}$ is +1 on bosons and −1 on fermions. We want to show that,

$$\mathrm{Tr}(-)^{N_F} = 0,$$

If the trace is taken over any finite-dimensional representation. $(-)^{N_F}$ anticommutes with Q. Using the cyclicity of the trace, one has,

$$0 = \mathrm{Tr}\left(-Q_\alpha (-)^{N_F} \bar{Q}_{\dot\beta} + (-)^{N_F} \bar{Q}_{\dot\beta} Q_\alpha \right) = \mathrm{Tr}\left((-)^{N_F} \left\{ Q_\alpha, \bar{Q}_{\dot\beta} \right\} \right)$$

$$= 2\sigma^\mu_{\alpha\dot\beta} \mathrm{Tr}\left((-)^{N_F} P_\mu \right).$$

Choosing any non-vanishing momentum P_μ gives the desired result.

Massless Supermultiplets

We will first assume that all central charges Z^{IJ} vanish. Below we will see that for massless representations this is necessarily a consequence of the positivity of the Hilbert space. Then all Q^I_α anticommute among themselves, and so do the $\bar{Q}^J_{\dot\beta}$. Since $P^2 = 0$ we choose a reference frame where $P_\mu = E(1, 0, 0, 1)$ so that $\sigma^\mu P_\mu = \begin{pmatrix} 0 & 0 \\ 0 & 2E \end{pmatrix}$ and thus,

$$\left\{Q^I_\alpha, \bar{Q}^J_{\dot\beta}\right\} = \begin{pmatrix} 0 & 0 \\ 0 & 4E \end{pmatrix}_{\alpha\dot\beta} \delta^{IJ}.$$

In particular, $\left\{Q^I_1, \bar{Q}^J_{\dot 1}\right\} = 0, \forall\, I, J$. On a positive definite Hilbert space we must then set $Q^I_1 = \bar{Q}^J_{\dot 1} = 0, \forall\, I, J$. The argument is similar to the one above:

$$0 = \left\langle \Phi \left| \left\{Q^I_1, \bar{Q}^I_{\dot 1}\right\} \right| \Phi \right\rangle = \left\| Q^I_1 |\Phi\rangle \right\|^2 + \left\| \bar{Q}^I_{\dot 1} |\Phi\rangle \right\|^2 \Rightarrow Q^I_1 = \bar{Q}^J_{\dot 1} = 0.$$

Thus we are left with only the Q^I_2 and $\bar{Q}^J_{\dot 2}$, i.e. N of the initial 2N fermionic generators. If we define,

$$a_I = \frac{1}{\sqrt{4E}} Q^I_2, \; a^\dagger_I = \frac{1}{\sqrt{4E}} \bar{Q}^I_{\dot 2},$$

the a_I and $a^\dagger_I$ are anticommuting annihilation and creation operators:

$$\left\{a_I, a^\dagger_J\right\} = \delta_{IJ}, \quad \left\{a_I, a_J\right\} = \left\{a^\dagger_I, a^\dagger_J\right\} = 0.$$

One then chooses a "vacuum state", i.e. a state annihilated by all the a_I. Such a state will carry some irreducible representation of the Poincaré algebra, i.e. in addition to its zero mass it is characterised by some helicity λ_0. We denote this state as $|\lambda_0\rangle$. From the commutators of Q^I_2 and $\bar{Q}^J_{\dot 2}$ with the helicity operator which in the present frame is $J_3 = M_{12}$ one sees that Q^I_2 lowers the helicity by one half and $\bar{Q}^J_{\dot 2}$ rises it by one half. (For simplicity, we suppose here that the state $|\lambda_0\rangle$ transforms as a singlet under the SU(N) that acts on the indices I, J. One could easily drop this restriction.) The supermultiplet then is of the form,

$$\begin{aligned} &|\lambda_0\rangle \\ a^\dagger_I |\lambda_0\rangle &= \left|\lambda_0 + \frac{1}{2}\right\rangle_I \\ a^\dagger_I a^\dagger_J |\lambda_0\rangle &= |\lambda_0 + 1\rangle_{IJ} \\ &\dots \end{aligned}$$

$$a_1^\dagger a_2^\dagger \dots a_N^\dagger |\lambda_0\rangle = \left|\lambda_0 + \frac{N}{2}\right\rangle.$$

Due to the antisymmetry in I, J, . . . there are $\binom{N}{k}$ states with helicity $\lambda = \lambda_0 + \frac{k}{2}$, k = 0, 1, . . .N. Summing the binomial coefficients gives a total of 2^N states with 2^{N-1} having integer helicity (bosons) and 2N−1 having half-integer helicity (fermions). In general, in such a supermultiplet, except if $\lambda_0 = -\frac{N}{4}$, the helicities will not be distributed symmetrically about zero. Such supermultiplets cannot be invariant under CPT, since CPT flips the sign of the helicity. To satisfy CPT one then need to double these multiplets by adding their CPT conjugate with opposite helicities and opposite quantum numbers. For unextended susy, N = 1, each massless supermultiplet only contains two states $|\lambda_0\rangle$ and $\left|\lambda_0 + \frac{1}{2}\right\rangle$. We denote these multiplets by $\left(\lambda_0, \lambda_0 + \frac{1}{2}\right)$. They can never be CPT self-conjugate and one needs to double them. Thus one arrives at the following massless N = 1 multiplets:

- The chiral multiplet consists of $\left(0, \frac{1}{2}\right)$ and its CPT conjugate $\left(-\frac{1}{2}, 0\right)$, corresponding to a Weyl fermion and a complex scalar.
- The vector multiplet consists of $\left(\frac{1}{2}, 1\right)$ plus $\left(-1, -\frac{1}{2}\right)$, corresponding to a gauge boson (massless vector) and a Weyl fermion, both necessarily in the adjoint representation of the gauge group.
- The gravitino multiplet contains $\left(1, \frac{3}{2}\right)$ and $\left(-\frac{3}{2}, -1\right)$ i.e. a gravitino and a gauge boson.
- The graviton multiplet contains $\left(\frac{3}{2}, 2\right)$ and $\left(-2, -\frac{3}{2}\right)$, corresponding to the graviton and the gravitino.

Since we so not want helicities larger than two, we must stop here. Also the gravitino should be present only in a theory with gravity, so if N = 1 it must only occur once and then in the gravity multiplet. Hence the gravitino multiplet cannot appear in unextended susy. However it does appear in extended susy when decomposing larger multiplets into N = 1 multiplets.

For N = 2 a supermultiplet contains $\left(\lambda_0, \lambda_0 + \frac{1}{2}, \lambda_0 + \frac{1}{2}, \lambda_0 + 1\right)$. Restricting ourselves to the cases where the helicity does not exceed one, we have two possibilities.

- The N = 2 vector multiplet contains $\left(0,\frac{1}{2},\frac{1}{2},1\right)$ and its CPT conjugate $\left(-1,-\frac{1}{2},-\frac{1}{2},0\right)$, corresponding to a vector (gauge boson), two Weyl fermions and a complex scalar, again all in the adjoint representation of the gauge group. In terms of N = 1 representations this is a vector and a chiral N = 1 multiplet.
- The hypermultiplet: If $\lambda_0=-\frac{1}{2}$ we get $\left(-\frac{1}{2},0,0,\frac{1}{2}\right)$. This may or not be CPT self-conjugate. If it is, it is called a half-hypermultiplet. If it is not we have to add its CPT conjugate to get a (full) hypermultiplet $2\times\left(-\frac{1}{2},0,0,\frac{1}{2}\right)$.

  For N = 4, restricting again to helicities not exceeding one, there is a single N = 4 multiplet which always is CPT self-conjugate. It is $\left(-1,4\times\left(-\frac{1}{2}\right),6\times 0,\ 4\times\frac{1}{2},1\right)$ containing a vector 4 Weyl fermions and 3 complex scalars. In terms of N = 2 multiplets it is just the sum of the N = 2 vector multiplet and a hypermultiplet, however now all transforming in the adjoint of the gauge group.

Massive Supermultiplets

We now consider the case $P^2 > 0$ and a priori arbitrary central charges Z^{IJ}. Going to the rest frame $P_\mu=(m,\ 0,\ 0,\ 0)$, the susy algebra becomes,

$$\left\{Q^I_\alpha,\ \left(Q^J_\beta\right)^\dagger\right\}=2m\delta_{\alpha\beta}\delta^{IJ}$$

$$\left\{Q^I_\alpha,Q^J_\beta\right\}=\epsilon_{\alpha\beta}\ Z^{IJ}$$

$$\left\{\left(Q^I_\alpha\right)^\dagger,\left(Q^J_\beta\right)^\dagger\right\}=\epsilon_{\alpha\beta}\left(Z^{IJ}\right)^*.$$

By an appropriate U(N) rotation among the Q^I the antisymmetric matrix of central charges can be brought into standard form:

$$Z^{IJ}=\begin{pmatrix} 0 & q_1 & 0 & 0 & \\ -q_1 & 0 & 0 & 0 & \cdots \\ 0 & 0 & 0 & q_2 & \\ 0 & 0 & -q_2 & 0 & \\ & \vdots & & & \end{pmatrix}$$

with all $q_n \geq 0$, $n = 1,\ldots\frac{N}{2}$. We assume that N is even, since otherwise there is an extra zero eigenvalue of the Z-matrix which can be handled trivially. It follows that if we let,

$$a_\alpha^1 = \frac{1}{\sqrt{2}}\left(Q_\alpha^1 + \epsilon_{\alpha\beta}\left(Q_\beta^2\right)^\dagger\right)$$
$$b_\alpha^1 = \frac{1}{\sqrt{2}}\left(Q_\alpha^1 - \epsilon_{\alpha\beta}\left(Q_\beta^2\right)^\dagger\right)$$
$$a_\alpha^2 = \frac{1}{\sqrt{2}}\left(Q_\alpha^3 + \epsilon_{\alpha\beta}\left(Q_\beta^4\right)^\dagger\right)$$
$$b_\alpha^2 = \frac{1}{\sqrt{2}}\left(Q_\alpha^3 - \epsilon_{\alpha\beta}\left(Q_\beta^4\right)^\dagger\right)$$
$$\vdots$$

then the a_α^r and b_α^r, $r = 1, \ldots \frac{N}{2}$ and their hermitian conjugates satisfy the following algebra of harmonic oscillators,

$$\left\{a_\alpha^r, \left(a_\beta^s\right)^\dagger\right\} = \left(2m - q_r\right)\delta_{rs}\delta_{\alpha\beta}$$
$$\left\{b_\alpha^r, \left(b_\beta^s\right)^\dagger\right\} = \left(2m + q_r\right)\delta_{rs}\delta_{\alpha\beta}$$
$$\left\{a_\alpha^r, \left(b_\beta^s\right)^\dagger\right\} = \left\{a_\alpha^r, a_\beta^s\right\} = \ldots = 0$$

Clearly, positivity of the Hilbert space requires that, $2m \geq |q_n|$, for all n. If some of the q_n saturate the bound, i.e. are equal to m, then the corresponding operators must be set to zero, as we did in the massless case with the Q_1^I. Clearly, in the massless case the bound becomes $0 \geq |q_n|$ and thus $q_n = 0$ always. There cannot be central charges in the massless case and the bound is always saturated, thus only exactly have of the fermionic generators survive.

In the more general massive case, if all $|q_n|$ are strictly less than 2m we have a total of 2N harmonic oscillators. Then starting from a state of minimal "helicity" (i.e. z component of the angular momentum) λ_0 annihilated by all a_α^n and b_β^n, application of the creation operators yields a total of 2^{2N} states with helicities ranging from λ_0 to $\lambda_0 + N$. For N = 1 this yields four states, again labeled by their helicities (or rather the z component of the angular momentum), as $\left(-\frac{1}{2}, 0, 0, \frac{1}{2}\right)$ (which is the same as the CPT extended massless multiplet) or $\left(-1, \frac{1}{2}, \frac{1}{2}, 0\right)$ to which we must add the CPT conjugate $\left(1, \frac{1}{2}, \frac{1}{2}, 0\right)$. The latter are the same states as a massless vector plus a massless chiral multiplet and can be obtained from them via a Higgs mechanism. In terms of massive representations this is a vector (3 dofs) a Dirac fermion (4 dofs) and a single real scalar (1 dof).

For N = 2 we already have 16 states with helicities ranging at least from −1 to 1. Such a massive N = 2 multiplet can be viewed as the union of a massless N = 2 vector and hypermultiplet. A generic massive N = 4 multiplet contains $2^8 = 256$ states including at least a helicity ± 2. Thus such a theory must include a massive spin two particle which is not believed to be possible in quantum field theory.

If $k < \frac{N}{2}$ of the q_n are equal to 2m then we only have 2N − 2k oscillators, and the super-multiplets will only contain $2^{2(N-k)}$ states. They are called short multiplets or BPS multiplets. If all q_n equal 2m, i.e. $k = \frac{N}{2}$ we get the shortest multiplets with only 2^N states, exactly as in the massless case. These BPS multiplets are also called ultrashort, and are completely analogous to the massless multiplets.

Superspace and Superfields

Since we want to construct supersymmetric quantum field theories, we have to find representations of the susy algebra on fields. A convenient and compact way to do this is to introduce superspace and superfields, i.e. fields defined on superspace. This is particularly simple for unextended susy, so we will restrict here to N = 1 superspace and superfields. Then we have two plus two susy generators Q_α and $\bar{Q}_{\dot{\alpha}}$, as well as four generators P_μ of space-time translations. The idea then is to enlarge space-time labelled by the coordinates x^μ by adding two plus two anticommuting Grassmannian coordinates θ_α and $\bar{\theta}_{\dot{\alpha}}$. Thus coordinates on superspace are $\left(x^\mu, \theta_\alpha, \bar{\theta}_{\dot{\alpha}}\right)$. Rather than elaborating on the meaning of such a space we will simply use it as a very efficient recepee to perform calculations in supersymmetric theories.

Superspace

As already said, we restrict here to N = 1. The "odd" superspace coordinates θ_α and $\bar{\theta}_{\dot{\alpha}}$ just behave as constant (x^μ independent) spinors. Recall that as all spinors they anticommute among themselves, i.e. $\theta^1\theta^2 = -\theta^2\theta^1$, and idem for the $\bar{\theta}^{\dot{\alpha}}$. Spinor indices in bilinears are contracted acording to the usual rule, i.e. $\theta\theta = \theta^\alpha\theta_\alpha = -2\theta^1\theta^2 = +2\theta_2\theta_1 = -2\theta_1\theta_2$, and $\overline{\theta\theta} = \bar{\theta}_{\dot{\alpha}}\bar{\theta}^{\dot{\alpha}} = 2\bar{\theta}_1\bar{\theta}_2 = \ldots$. One can then easily prove the following useful identities:

$$\theta^\alpha\theta^\beta = -\frac{1}{2}\epsilon^{\alpha\beta}\,\theta\theta\,,\ \bar{\theta}^{\dot{\alpha}}\bar{\theta}^{\dot{\beta}} = \frac{1}{2}\epsilon^{\dot{\alpha}\dot{\beta}}\,\overline{\theta\theta},$$

$$\theta_\alpha\theta_\beta = \frac{1}{2}\epsilon_{\alpha\beta}\,\theta\theta\,,\ \bar{\theta}_{\dot{\alpha}}\bar{\theta}_{\dot{\beta}} = -\frac{1}{2}\epsilon_{\dot{\alpha}\dot{\beta}}\,\overline{\theta\theta},$$

$$\theta\sigma^\mu\bar{\theta}\theta\sigma^\nu\bar{\theta} = \frac{1}{2}\theta\theta\,\overline{\theta\theta}\,g^{\mu\nu},\ \theta\psi\theta\chi = -\frac{1}{2}\theta\theta\psi\chi.$$

Derivatives in θ and $\bar{\theta}$ are defined in an obvious way as $\frac{\partial}{\theta^{\alpha}}\theta^{\beta}=\delta_{\alpha}^{\beta}$ and $\frac{\partial}{\bar{\theta}^{\dot{\alpha}}}\bar{\theta}^{\dot{\beta}}=\delta_{\dot{\alpha}}^{\dot{\beta}}$. Since the θ's anticommute, any product involving more than two θ's or more than two θ's vanishes. Hence an arbitrary (scalar) function on superspace, i.e. a superfield, can always be expanded as,

$$F(x,\theta,\bar{\theta})=f(x)+\theta\psi(x)+\bar{\theta}\bar{\chi}(x)+\theta\theta\, m(x)+\bar{\theta}\bar{\theta}n(x)$$
$$+\theta\sigma^{\mu}\bar{\theta}v_{\mu}(x)+\theta\theta\;\bar{\theta}\bar{\lambda}(x)+\bar{\theta}\bar{\theta}\;\theta\rho(x)+\theta\theta\;\bar{\theta}\bar{\theta}\,d(x).$$

If F carries extra vector indices then so do the fomponent fields $f,\psi,\ \ldots$

Integration on superspace is defined for a single Grassmannian variable, say θ^{1} as $\int d\theta^{1}\left(a+\theta^{1}b\right)=b$ so that $\int d\theta^{1}d\theta^{2}\theta^{2}\theta^{1}=1$. Then since $\theta\theta=2\theta^{2}\theta^{1}$ and $\bar{\theta}\bar{\theta}=2\bar{\theta}^{1}\bar{\theta}^{2}$ we define $d^{2}\theta=\frac{1}{2}d\theta^{1}d\theta^{2}$ and $d^{2}\bar{\theta}=\frac{1}{2}d\bar{\theta}^{2}d\bar{\theta}^{1}=\left[d^{2}\theta\right]^{\dagger}$ so that, $\int d^{2}\theta\;\theta\theta=\int d^{2}\bar{\theta}\;\bar{\theta}\bar{\theta}=1$.

It is easy to check that, $\int d^{2}\theta=\frac{1}{4}\epsilon^{\alpha\beta}\frac{\partial}{\partial\theta^{\alpha}}\frac{\partial}{\partial\theta^{\beta}}$, $\int d^{2}\bar{\theta}=-\frac{1}{4}\epsilon^{\dot{\alpha}\dot{\beta}}\frac{\partial}{\partial\bar{\theta}^{\dot{\alpha}}}\frac{\partial}{\partial\bar{\theta}^{\dot{\beta}}}$. Clearly one also has, $\int d^{2}\theta d^{2}\bar{\theta}\;\theta\theta\bar{\theta}\bar{\theta}=1$. With these definitions it is easy to see that one has the hermiticity property, $\left(\frac{\partial}{\theta^{\alpha}}\right)^{\dagger}=+\frac{\partial}{\bar{\theta}^{\dot{\alpha}}}$ with $\alpha\equiv\dot{\alpha}$. The plus sign rather than a minus sign as one would expect from $\left(\partial_{\mu}\right)^{\dagger}=-\partial_{\mu}$.

We now want to realise the susy generators Q_{α} and their hermitian conjugates $\bar{Q}_{\dot{\alpha}}=\left(Q_{\alpha}\right)^{\dagger}$ as differential operators on superspace. We want that $i\epsilon^{\alpha}Q_{\alpha}$ generates a translation in θ^{α} by a constant infinitesimal spinor ϵ^{α} plus some translation in x^{μ}. The latter space-time translation is determined by the susy algebra since the commutator of two such susy transformations is a translation in space-time. Thus we want, $(1+i\epsilon Q)F(x,\theta,\theta)=F\left(x+\delta x,\theta+\epsilon,\bar{\theta}\right)$.

Hence $iQ_{\alpha}=\frac{\partial}{\partial\theta^{\alpha}}+\ldots$ where $+\ldots$ must be of the form $c\left(\sigma^{\mu}\bar{\theta}\right)_{\alpha}P_{\mu}=-ic\left(\sigma^{\mu}\bar{\theta}\right)_{\alpha}\partial_{\mu}$ with some constant c to be determined. We arrive at the ansatz,

$$Q_{\alpha}=-i\left(\frac{\partial}{\partial\theta^{\alpha}}-ic\left(\sigma^{\mu}\bar{\theta}\right)_{\alpha}\partial_{\mu}\right).$$

Then the hermitian conjugate is,

$$\bar{Q}_{\dot{\alpha}}=i\left(\frac{\partial}{\partial\bar{\theta}^{\dot{\alpha}}}-ic^{*}\left(\theta\sigma^{\mu}\right)_{\dot{\alpha}}\partial_{\mu}\right),$$

and they satisfy the susy algebra, in particular,

$$\left\{Q_\alpha, Q_{\dot\beta}\right\} = 2\sigma^\mu_{\alpha\dot\beta} P_\mu = -2i\sigma^\mu_{\alpha\dot\beta}\partial_\mu$$

if Re c = 1. We choose c = 1 so that,

$$Q_\alpha = -i\frac{\partial}{\partial\theta^\alpha} - \sigma^\mu_{\alpha\dot\beta}\bar\theta^{\dot\beta}\partial_\mu$$

$$\bar Q_{\dot\alpha} = i\frac{\partial}{\partial\bar\theta^{\dot\alpha}} + \theta^\beta\sigma^\mu_{\beta\dot\alpha}\partial_\mu.$$

We can now give the action on the superfield F and determine δx:

$$\left(1 + i\in Q + i\bar\in\bar Q\right)F\left(x^\mu, \theta^\alpha, \bar\theta^{\dot\beta}\right) = F\left(x^\mu - i\in\sigma^\mu\bar\theta + i\theta\sigma^\mu\bar\in, \theta^\alpha + \in^\alpha, \bar\theta^{\dot\beta} + \bar\in^{\dot\beta}\right)$$

and the susy variation of a superfield is of course defined as,

$$\delta_{\in,\bar\in}F = \left(i\in Q + i\bar\in\bar Q\right)F.$$

Since a general superfield contains too many component fields to correspond to an irreducible representation of N = 1 susy, it will be very useful to impose susy invariant condition to lower the number of components. To do this, we first find covariant derivatives D_α and $\bar D_{\dot\alpha}$ that anticommute with the susy generators Q and $\bar Q$. Then $\delta_{\in,\bar\in}(D_\alpha F) = D_\alpha(\delta_{\in,\bar\in}F)$ and idem for $\bar D_{\dot\alpha}$. It follows that $D_\alpha F = 0$ or $\bar D_{\dot\alpha}F = 0$ are susy invariant constraints one may impose to reduce the number of components in a superfield. One finds,

$$D_\alpha = \frac{\partial}{\partial\theta^\alpha} + i\sigma^\mu_{\alpha\dot\beta}\bar\theta^{\dot\beta}\partial_\mu$$

$$\bar D_{\dot\alpha} = \frac{\partial}{\partial\bar\theta^{\dot\alpha}} + i\theta^\beta\sigma^\mu_{\beta\dot\alpha}\partial_\mu$$

where $\bar D_{\dot\alpha} = (D_\alpha)^\dagger$ and,

$$\left\{D_\alpha, \bar D_{\dot\beta}\right\} = 2i\sigma^\mu_{\alpha\dot\beta}\partial_\mu, \left\{D_\alpha, D_\beta\right\} = \left\{\bar D_{\dot\alpha}, \bar D_{\dot\beta}\right\} = 0$$

$$\left\{D_\alpha, Q_\beta\right\} = \left\{\bar D_{\dot\alpha}, Q_\beta\right\} = \left\{D_\alpha, \bar Q_{\dot\beta}\right\} = \left\{\bar D_{\dot\alpha}, \bar Q_{\dot\beta}\right\} = 0.$$

Chiral Superfields

A chiral superfield ϕ is defined by the condition, $\bar D_{\dot\alpha}\phi = 0$ and an anti-chiral one $\bar\phi$ by $D_\alpha\bar\phi = 0$. This is easily solved by observing that,

$$D_\alpha \bar{\theta} = \bar{D}_{\dot{\alpha}} \theta = D_\alpha \bar{y}^\mu = \bar{D}_{\dot{\alpha}} y^\mu = 0,$$
$$y^\mu = x^\mu + i\theta\sigma^\mu\bar{\theta}, \quad \bar{y}^\mu = x^\mu - i\theta\sigma^\mu\bar{\theta}.$$

Hence ϕ depends only on θ and y^μ (i.e. all $\bar{\theta}$ dependence is through y^μ) and $\bar{\phi}$ only on $\bar{\theta}$ and $\bar{y}^\mu$. Concentrating on ϕ we have the component expansion,

$$\phi(y, \theta) = z(y) + \sqrt{2}\theta\psi(y) - \theta\theta f(y)$$

or Taylor expanding in terms of x, θ and $\bar{\theta}$:

$$\phi(y, \theta) = z(x) + \sqrt{2}\theta\psi(x) + i\theta\sigma^\mu\bar{\theta}\partial_\mu z(x) - \theta\theta f(x)$$
$$-\frac{i}{\sqrt{2}}\theta\theta\partial_\mu\psi(x)\sigma^\mu\bar{\theta} - \frac{1}{4}\theta\theta\bar{\theta}\bar{\theta}\partial^2 z(x).$$

Physically, such a chiral superfield describes one complex scalar z and one Weyl fermion ψ. The field f will turn out to be an auxiliary field. For $\bar{\phi}$ we similarly have,

$$\phi(y, \bar{\theta}) = \bar{z}(\bar{y}) + \sqrt{2}\bar{\theta}\bar{\psi}(\bar{y}) - \bar{\theta}\bar{\theta}\bar{f}(\bar{y})$$
$$= \bar{z}(x) + \sqrt{2}\bar{\theta}\bar{\psi}(x) - i\theta\sigma^\mu\bar{\theta}\partial_\mu\bar{z}(x) - \bar{\theta}\bar{\theta}\bar{f}(x)$$
$$+\frac{i}{\sqrt{2}}\bar{\theta}\bar{\theta}\theta\sigma^\mu\partial_\mu\bar{\psi}(x) - \frac{1}{4}\theta\theta\bar{\theta}\bar{\theta}\partial^2\bar{z}(x).$$

Finally, let us find the explicit susy variations of the component fields as it results from $\delta_{\epsilon,\bar{\epsilon}}F = (i\epsilon Q + i\bar{\epsilon}\bar{Q})F$: First, for chiral superfields it is useful to change variables from $x^\mu, \theta, \bar{\theta}$ to $y^\mu, \theta, \bar{\theta}$. Then,

$$Q_\alpha = -i\frac{\partial}{\partial\theta^\alpha}, \quad \bar{Q}_{\dot{\alpha}} = i\frac{\partial}{\partial\bar{\theta}^{\dot{\alpha}}} + 2\theta^\beta\sigma^\mu_{\beta\dot{\alpha}}\frac{\partial}{\partial y^\mu},$$

so that,

$$\delta\phi(y, \theta) \equiv (i\epsilon Q + i\bar{\epsilon}\bar{Q})\phi(y,\theta) = \left(\epsilon^\alpha\frac{\partial}{\partial\theta^\alpha} + 2i\theta\sigma^\mu\bar{\epsilon}\frac{\partial}{\partial y^\mu}\right)\phi(y,\theta)$$
$$= \sqrt{2}\epsilon\psi - 2\epsilon\theta f + 2i\theta\sigma^\mu\bar{\epsilon}\left(\partial_\mu z + \sqrt{2}\theta\partial_\mu\psi\right)$$
$$= \sqrt{2}\epsilon\psi + \sqrt{2}\theta\left(-\sqrt{2}\epsilon f + \sqrt{2}i\sigma^\mu\bar{\epsilon}\partial_\mu z\right) - \theta\theta\left(-i\sqrt{2}\bar{\epsilon}\bar{\sigma}^\mu\partial_\mu\psi\right).$$

Thus we read the susy transformations of the component fields:

$$\delta z = \sqrt{2}\epsilon\psi$$
$$\delta\psi = \sqrt{2}i\partial_\mu z\sigma^\mu\bar{\epsilon} - \sqrt{2}f\epsilon$$

$$\delta f = \sqrt{2} i \partial_\mu \psi \sigma^\mu \bar{\epsilon}.$$

The factors of $\sqrt{2}$ do appear because of our normalisations of the fields and the definition of $\delta\phi$. If desired, they could be absorbed by a rescaling of ϵ and $\bar{\epsilon}$.

Susy Invariant Actions

To construct susy invariant actions we now only need to make a few observations. First, products of superfields are of course superfields. Also, products of (anti) chiral superfields are still (anti) chiral superfields. Typically, one will have a superpotential $W(\phi)$ which is again chiral. This W may depend on several different ϕ_i. Using the y and θ variables one easily Taylor expands,

$$W(\phi) = W(z(y)) + \sqrt{2}\frac{\partial W}{\partial z_i}\theta\psi_i(y)$$
$$-\theta\theta\left(\frac{\partial W}{\partial z_i}f_i(y) + \frac{1}{2}\frac{\partial^2 W}{\partial z_i \partial z_j}\psi_i(y)\psi_j(y)\right),$$

where it is understood that $\partial W/\partial z$ and $\partial^2 W/\partial z \partial z$ are evaluated at z(y). The second and important observation is that any Lagrangian of the form,

$$\int d^2\theta d^2\bar{\theta}\, F(x,\theta,\bar{\theta}) + \int d^2\theta\, W(\phi) + \int d^2\bar{\theta}\, [W(\phi)]^\dagger,$$

is automatically susy invariant, i.e. it transforms at most by a total derivative in space-time. The proof is very simple. The susy variation of any superfield is given by $\delta_{\epsilon,\bar{\epsilon}}F = (i\epsilon Q + i\bar{\epsilon}\bar{Q})F$ and, since the ϵ and $\bar{\epsilon}$ are constant spinors and the Q and $\bar{Q}$ are differential operators in superspace, it is again a total derivative in all of superspace:

$$\delta F = \frac{\partial}{\partial\theta^\alpha}(-\epsilon^\alpha F) + \frac{\partial}{\partial\bar{\theta}^{\dot{\alpha}}}(-\bar{\epsilon}^{\dot{\alpha}} F) + \frac{\partial}{\partial x^\mu}\left[-i(\epsilon\sigma^\mu\bar{\theta} - \theta\sigma^\mu\bar{\epsilon})F\right].$$

Integration $\int d^2\theta d^2\bar{\theta}$ only leaves the last term which is a total space-time derivative as claimed. If now F is a chiral superfield like ϕ or $W(\phi)$ one changes variables to θ and y and one has,

$$\delta\phi = \frac{\partial}{\partial\theta^\alpha}(-\epsilon^\alpha\phi(y,\theta)) + \frac{\partial}{\partial y^\mu}\left[-i(\epsilon\sigma^\mu\bar{\theta} - \theta\sigma^\mu\bar{\epsilon})\phi(y,\theta)\right].$$

Integrating $\int d^2\theta$ again only leaves the last term which becomes $\frac{\partial}{\partial x^\mu}[\ldots]$ and is a total derivative in space-time. The analogous result holds for an anti-chiral superfield $\overline{W}(\bar{\phi}) = [W(\phi)]^\dagger$ and integration $\int d^2\bar{\theta}$. This proves the supersymmetry of the action resulting from the space-time integral of the Lagrangian $\int d^2\theta d^2\bar{\theta}\, F(x,\theta,\bar{\theta}) + \int d^2\theta\, W(\phi) + \int d^2\bar{\theta}\, [W(\phi)]^{\dagger}$.

The terms $\int d^2\theta W(\phi) + \text{h.c.}$ in the Lagrangian have the form of a potential. The kinetic terms must be provided by the term $\int d^2\theta d^2\bar{\theta}\, F$. The simplest choice is $F = \phi^{\dagger}\phi$. This is neither chiral nor anti chiral but real. To compute $\phi^{\dagger}\phi$ one must first expand the y^{μ} in terms of x^{μ}. We only need the terms $\sim \theta\theta\bar{\theta}\bar{\theta}$, called the D-term:

$$\phi^{\dagger}\phi\Big|_{\theta\theta\bar{\theta}\bar{\theta}} = -\frac{1}{4}z^{\dagger}d^2z - \frac{1}{4}d^2z^{\dagger}z + \frac{1}{2}\partial_{\mu}z^{\dagger}\partial^{\mu}z + f^{\dagger}f + \frac{i}{2}\partial_{\mu}\psi\sigma^{\mu}\bar{\psi} - \frac{i}{2}\psi\sigma^{\mu}\partial_{\mu}\bar{\psi}$$

$$= \partial_{\mu}z^{\dagger}\partial^{\mu}z + \frac{i}{2}\left(\partial_{\mu}\psi\sigma^{\mu}\bar{\psi} - \psi\sigma^{\mu}\partial_{\mu}\bar{\psi}\right) + f^{\dagger}f + \text{total derivative.}$$

Then, $S = \int d^4x d^2\theta d^2\bar{\theta}\, \bar{\phi}_i^{\dagger}\, \phi_i + \int d^4x d^2\theta W(\phi_i) + \text{h.c.}$ yields,

$$S = \int d^4x \left[\left|\partial_{\mu}z_i\right|^2 - i\psi_i\sigma^{\mu}\partial_{\mu}\bar{\psi}_i + f_i^{\dagger}f_i - \frac{\partial W}{\partial z_i}f_i + \text{h.c.} - \frac{1}{2}\frac{\partial^2 W}{\partial z_i \partial z_j}\psi_i\psi_j + \text{h.c.} \right].$$

More generally, one can replace $\phi_i^{\dagger}\, \phi_i$ by a (real) Kähler potential $K(\phi_i^{\dagger}\, \phi_j)$. This leads to the non-linear σ-model discussed later. In any case, the f_i have no kinetic term and hence are auxiliary fields. They should be eliminated by substituting their algebraic equations of motion,

$$f_i^{\dagger} = \left(\frac{\partial W}{\partial z_i}\right)$$

into the action, leading to,

$$S = \int d^4x \left[\left|\partial_{\mu}z_i\right|^2 - i\psi_i\sigma^{\mu}\partial_{\mu}\bar{\psi}_i - \left|\frac{\partial W}{\partial z_i}\right|^2 - \frac{1}{2}\frac{\partial^2 W}{\partial z_i \partial z_j}\psi_i\psi_j - \frac{1}{2}\left(\frac{\partial^2 W}{\partial z_i \partial z_j}\right)^{\dagger}\bar{\psi}_i\bar{\psi}_j \right].$$

We see that the scalar potential V is determined in terms of the superpotential W as,

$$V = \sum_i \left|\frac{\partial W}{\partial z_i}\right|^2.$$

To illustrate this model, consider the simplest case of a single chiral superfield ϕ and a cubic superpotential $W(\phi) = \frac{m}{2}\phi^2 + \frac{g}{3}\phi^3$. Then $\frac{\partial W}{\partial z} = m\phi + g\phi^2$ and the action becomes,

$$S_{WZ} = \int d^4x \Big[\left|\partial_{\mu}z\right|^2 - i\psi\sigma^{\mu}\partial_{\mu}\bar{\psi} - m^2|z|^2 - \frac{m}{2}(\psi\psi + \bar{\psi}\bar{\psi})$$
$$- mg\left(z^{\dagger}z^2 + \left(z^{\dagger}\right)^2 z\right) - g^2|z|^4 + g\left(z\psi\psi + z^{\dagger}\bar{\psi}\bar{\psi}\right) \Big].$$

The Yukawa interactions appear with a coupling constant g that is related by susy to the bosonic coupling constants mg and g^2.

Vector Superfields

The N = 1 supermultiplet of next higher spin is the vector multiplet. The corresponding superfield $V(x,\theta,\bar\theta)$ is real and has the expansion,

$$V(x,\theta,\bar\theta)=C+i\theta\chi-i\bar\theta\bar\chi+\theta\sigma^\mu\bar\theta v_\mu+\frac{i}{2}\theta\theta(M+iN)-\frac{i}{2}\bar\theta\bar\theta(M-iN)$$
$$+i\theta\theta\bar\theta\left(\bar\lambda+\frac{i}{2}\bar\sigma^\mu\partial_\mu\chi\right)-i\bar\theta\bar\theta\theta\left(\lambda-\frac{i}{2}\sigma^\mu\partial_\mu\bar\chi\right)$$
$$+\frac{i}{2}\theta\theta\bar\theta\bar\theta\left(D-\frac{1}{2}\partial^2C\right)$$

where all component fields only depend on x^μ. There are 8 bosonic components (C, D, M, N, v_μ) and 8 fermionic components (χ, λ). These are too many components to describe a single supermultiplet. We want to reduce their number by making use of the supersymmetric generalisation of a gauge transformation. The transformation,

$$V\rightarrow V+\phi+\phi^\dagger,$$

with ϕ a chiral superfield, implies the component transformation,

$$v_\mu\rightarrow v_\mu+\partial_\mu(2\text{Im}z)$$

which is an abelian gauge transformation. We conclude that $V\rightarrow V+\phi+\phi^\dagger$, is its desired supersymmetric generalisation. If this transformation $V\rightarrow V+\phi+\phi^\dagger$, is a symmetry (actually a gauge symmetry, as we just saw) of the theory then, by an appropriate choice of ϕ, one can transform away the components χ, C, M, N and one component of v_μ. This choice is called the Wess-Zumino gauge, and it reduces the vector superfield to,

$$V_{WZ}=\theta\sigma^\mu\bar\theta v_\mu(x)+i\theta\theta\,\bar\theta\bar\lambda(x)-i\bar\theta\bar\theta\,\theta\lambda(x)+\frac{1}{2}\theta\theta\bar\theta\bar\theta D(x).$$

Since each term contains at least one θ, the only non-vanishing power of V_{WZ} is,

$$V_{WZ}^2=\theta\sigma^\mu\bar\theta\,\theta\sigma^\nu\,\bar\theta\,v_\mu v_\nu=\frac{1}{2}\theta\theta\bar\theta\bar\theta v_\mu v^\mu \text{ and } V_{WZ}^n=0, n\geq 3.$$

To construct kinetic terms for the vector field v_μ one must act on V with the covariant derivatives D and $\bar D$. Define,

$$W_\alpha=-\frac{1}{4}\bar D\bar D D_\alpha V,\ \bar W_{\dot\alpha}=-\frac{1}{4}DD\bar D_{\dot\alpha}V.$$

(This is appropriate for abelian gauge theories and will be slightly generalized in the non-abelian case.) Since $D^3 = \overline{D}^3 = 0$, W_α is chiral and $\overline{W}_{\dot\alpha}$ antichiral. Furthermore it is clear that they behave as anticommuting Lorentz spinors. They are invariant under the transformation $V \to V + \phi + \phi^\dagger$, since,

$$W_\alpha \to W_\alpha - \frac{1}{4}\overline{D}\overline{D}D_\alpha\left(\phi + \phi^\dagger\right) = W_\alpha + \frac{1}{4}\overline{D}^{\dot\beta}\overline{D}_{\dot\beta}D_\alpha\phi$$

$$= W_\alpha + \frac{1}{4}\overline{D}^{\dot\beta}\left\{\overline{D}_{\dot\beta}, D_\alpha\right\}\phi = W_\alpha + \frac{i}{2}\sigma^\mu_{\alpha\dot\beta}\partial_\mu\overline{D}^{\dot\beta}\phi = W_\alpha$$

since $\overline{D}\phi = D\phi^\dagger = 0$. It is then easiest to use the WZ-gauge to compute W_α. To facilitate things further, change variables to y^μ, θ^α, $\overline{\theta}^{\dot\alpha}$ so that,

$$D_\alpha = \frac{\partial}{\partial\theta^\alpha} + 2i\sigma^\mu_{\alpha\dot\beta}\overline{\theta}^{\dot\beta}\frac{\partial}{\partial y^\mu}, \qquad \overline{D}_{\dot\alpha} = \frac{\partial}{\partial\overline{\theta}^{\dot\alpha}}$$

and write,

$$V_{WZ} = \theta\sigma^\mu\overline{\theta}v_\mu(y) + i\theta\theta\,\overline{\theta}\overline{\lambda}(y) - i\overline{\theta}\overline{\theta}\,\theta\lambda(y) + \frac{1}{2}\,\theta\theta\overline{\theta}\overline{\theta}\left(D(y) - i\partial_\mu v^\mu(y)\right).$$

Then, using $\sigma^\nu\overline{\sigma}^\mu - g^{\nu\mu} = 2\sigma^{\nu\mu}$, it is straightforward to find (all arguments are y^μ),

$$D_\alpha V_{WZ} = \left(\sigma^\mu\overline{\theta}\right)_\alpha v_\mu + 2i\theta_\alpha\overline{\theta}\overline{\lambda} - i\overline{\theta}\overline{\theta}\lambda_\alpha + \theta_\alpha\overline{\theta}\overline{\theta}D$$
$$+ 2i\left(\sigma^{\mu\nu}\theta\right)_\alpha\overline{\theta}\overline{\theta}\partial_\mu v_\nu + \theta\theta\overline{\theta}\overline{\theta}\left(\sigma^\mu\partial_\mu\overline{\lambda}\right)_\alpha$$

and then, using $\overline{D}\overline{D}\overline{\theta}\overline{\theta} = -4$,

$$W_\alpha = -i\lambda_\alpha(y) + \theta_\alpha D(y) + i\left(\sigma^{\mu\nu}\theta\right)_\alpha f_{\mu\nu}(y) + \theta\theta\left(\sigma^\mu\partial_\mu\overline{\lambda}(y)\right)_\alpha$$

with,

$$f_{\mu\nu} = \partial_\mu v_\nu - \partial_\nu v_\mu$$

being the abelian field strength associated with v_μ.

Since, W_α is a chiral superfield, $\int d^2\theta\, W^\alpha W_\alpha$ will be a susy invariant Lagrangian. To obtain its component expansion we need the $\theta\theta$-term (F-term) of $W^\alpha W_\alpha$:

$$W^\alpha W_\alpha\big|_{\theta\theta} = -2i\lambda\sigma^\mu\partial_\mu\overline{\lambda} + D^2 - \frac{1}{2}\left(\sigma^{\mu\nu}\right)^{\alpha\beta}\left(\sigma^{\rho\sigma}\right)_{\alpha\beta}f_{\mu\nu}f_{\rho\sigma},$$

where we used $\left(\sigma^{\mu\nu}\right)_{\alpha}^{\beta} = \mathrm{tr}\,\sigma^{\mu\nu} = 0$. Furthermore,

$$\left(\sigma^{\mu\nu}\right)^{\alpha\beta}\left(\sigma^{\rho\sigma}\right)_{\alpha\beta} = \frac{1}{2}\left(g^{\mu\rho}g^{\nu\sigma} - g^{\mu\sigma}g^{\nu\rho}\right) - \frac{i}{2}\epsilon^{\mu\nu\rho\sigma},$$

(with $\epsilon^{0123} = +1$) so that,

$$\int d^2\theta\, W^{\alpha}W_{\alpha} = -\frac{1}{2}f_{\mu\nu}f^{\mu\nu} - 2i\lambda\sigma^{\mu}\partial_{\mu}\bar{\lambda} + D^2 + \frac{i}{4}\epsilon^{\mu\nu\rho\sigma}f_{\mu\nu}f_{\rho\sigma}.$$

The first three terms are real while the last one is purely imaginary.

Supersymmetric Gauge Theories

All members of the vector multiplet (the gauge boson v_{μ} and the gaugino λ) necessarily are in the same representation of the gauge group, i.e. in the adjoint representation. The chiral fields can be in any representation of the gauge group, e.g. in the fundamental one.

Pure N = 1 Gauge Theory

We start with the vector multiplet $V(x,\theta,\bar{\theta}) = C + i\theta\chi - i\bar{\theta}\bar{\chi} + \theta\sigma^{\mu}\bar{\theta}v_{\mu} + \frac{i}{2}\theta\theta(M+iN)$ $-\frac{i}{2}\bar{\theta}\bar{\theta}(M-iN) + i\theta\theta\bar{\theta}\left(\bar{\lambda} + \frac{i}{2}\bar{\sigma}^{\mu}\partial_{\mu}\chi\right) - i\bar{\theta}\bar{\theta}\theta\left(\lambda - \frac{i}{2}\sigma^{\mu}\partial_{\mu}\bar{\chi}\right) + \frac{i}{2}\theta\theta\bar{\theta}\bar{\theta}\left(D - \frac{1}{2}\partial^2 C\right)$ with every component now in the adjoint representation of the gauge group G, i.e. $V \equiv V_a T^a, a = 1, \ldots \dim G$ where the T_a are the generators in the adjoint. The basic object then is e^V rather than V itself. The non-abelian generalisation of the transformation $V \to V + \phi + \phi^{\dagger}$, is now,

$$e^V \to e^{i\Lambda^{\dagger}}e^V e^{-i\Lambda} \Leftrightarrow e^{-V} \to e^{i\Lambda}e^{-V}e^{-i\Lambda^{\dagger}},$$

with Λ a chiral superfield. To first order in Λ this reproduces $V \to V + \phi + \phi^{\dagger}$, with $\phi = -i\Lambda$. We will construct an action such that this non-linear transformation is a (local) symmetry. This transformation can again be used to set χ, C, M, N and one component of v_{μ} to zero, resulting in the same component expansion $V_{WZ} = \theta\sigma^{\mu}\bar{\theta}v_{\mu}(x) + i\theta\theta\,\bar{\theta}\bar{\lambda}(x) - i\bar{\theta}\bar{\theta}\,\theta\lambda(x) + \frac{1}{2}\theta\theta\bar{\theta}\bar{\theta}D(x)$ of V in the Wess-Zumino gauge. From now on we adopt this WZ gauge. Then $V^n = 0, n \geq 3$. The same remains true if some D_{α} or $\bar{D}_{\dot{\alpha}}$ are inserted in the product, e.g. $V(D_{\alpha}V)V = 0$. One simply has,

$$e^V = 1 + V + \frac{1}{2}V^2.$$

The superfields W_α are now defined as,

$$W_\alpha = -\frac{1}{4}\overline{DD}\left(e^{-V}D_\alpha e^{V}\right),\ \overline{W}_{\dot\alpha} = +\frac{1}{4}DD\left(e^{V}\overline{D}_{\dot\alpha}e^{-V}\right),$$

which to first order in V reduces to the abelian definition $W_\alpha = -\frac{1}{4}\overline{DD}D_\alpha V$, $\overline{W}_{\dot\alpha} = -\frac{1}{4}DD\overline{D}_{\dot\alpha}V$. Under the transformation $e^V \rightarrow e^{i\Lambda^\dagger}e^V e^{-i\Lambda} \Leftrightarrow e^{-V} \rightarrow e^{i\Lambda}e^{-V}e^{-i\Lambda^\dagger}$ one then has,

$$W_\alpha \rightarrow -\frac{1}{4}\overline{DD}\left(e^{i\Lambda}e^{-V}e^{-i\Lambda^\dagger} D_\alpha \left(e^{i\Lambda^\dagger}e^{V}e^{-i\Lambda}\right)\right)$$

$$= -\frac{1}{4}\overline{DD}\left(e^{i\Lambda}e^{-V}\left(\left(D_\alpha e^{V}\right)e^{-i\Lambda} + e^{V}D_\alpha e^{-i\Lambda}\right)\right).$$

The second term is $-\frac{1}{4}\overline{DD}\left(e^{i\Lambda}D_\alpha e^{-i\Lambda}\right)$ and vanishes for the same reason as $\frac{1}{4}\overline{DD}D_\alpha\phi$ in $W_\alpha \rightarrow W_\alpha - \frac{1}{4}\overline{DD}D_\alpha\left(\phi+\phi^\dagger\right) = W_\alpha + \frac{1}{4}\overline{D}^{\dot\beta}\overline{D}_{\dot\beta}D_\alpha\phi = W_\alpha + \frac{1}{4}\overline{D}^{\dot\beta}\left\{\overline{D}_{\dot\beta},D_\alpha\right\}\phi = W_\alpha + \frac{i}{2}\sigma^\mu_{\alpha\dot\beta}\partial_\mu\overline{D}^{\dot\beta}\phi$ $= W_\alpha$. Thus,

$$W_\alpha \rightarrow -\frac{1}{4}e^{i\Lambda}\overline{DD}\left(e^{-V}D_\alpha e^{V}\right)e^{-i\Lambda} = e^{i\Lambda}W_\alpha e^{-i\Lambda},$$

i.e. W_α transforms covariantly under $e^V \rightarrow e^{i\Lambda^\dagger}e^V e^{-i\Lambda} \Leftrightarrow e^{-V} \rightarrow e^{i\Lambda}e^{-V}e^{-i\Lambda^\dagger}$. Similarly, one has,

$$\overline{W}_{\dot\alpha} \rightarrow e^{i\Lambda^\dagger}W_\alpha e^{-i\Lambda^\dagger}.$$

Next, we want to obtain the component expansion of W_α in WZ gauge. Inserting the expansion $e^V = 1 + V + \frac{1}{2}V^2$, into the definition $W_\alpha = -\frac{1}{4}\overline{DD}\left(e^{-V}D_\alpha e^{V}\right)$, $\overline{W}_{\dot\alpha} = +\frac{1}{4}DD\left(e^{V}\overline{D}_{\dot\alpha}e^{-V}\right)$, gives,

$$W_\alpha = -\frac{1}{4}\overline{DD}D_\alpha V + \frac{1}{8}\overline{DD}\left[V,D_\alpha V\right].$$

The first term is the same as in the abelian case while for the new term we have (all arguments are y_μ),

$$\left[V,D_\alpha V\right] = \overline{\theta\theta}\left(\sigma^{\nu\mu}\theta\right)_\alpha\left[v_\mu,v_\nu\right] + i\theta\theta\overline{\theta\theta}\sigma^\mu_{\alpha\dot\beta}\left[v_\mu, \overline{\lambda}^{\dot\beta}\right]$$

and then, using again $\overline{D}\overline{D}\overline{\theta}\overline{\theta} = -4$,

$$\frac{1}{8}\overline{D}\overline{D}[V, D_\alpha V] = \frac{1}{2}(\sigma^{\mu\nu}\theta)_\alpha [v_\mu, v_\nu] - \frac{i}{2}\theta\theta\sigma^\mu_{\alpha\dot\beta}[v_\mu, \overline{\lambda}^{\dot\beta}].$$

Adding this to $W_\alpha = -i\lambda_\alpha(y) + \theta_\alpha D(y) + i(\sigma^{\mu\nu}\theta)_\alpha f_{\mu\nu}(y) + \theta\theta(\sigma^\mu\partial_\mu\overline{\lambda}(y))_\alpha$ simply turns ordinary derivatives of the fields into gauge covariant derivatives and we finally obtain,

$$W_\alpha = -i\lambda_\alpha(y) + \theta_\alpha D(y) + i(\sigma^{\mu\nu}\theta)_\alpha F_{\mu\nu}(y) + \theta\theta(\sigma^\mu\partial_\mu\overline{\lambda}(y))_\alpha$$

where now,

$$F_{\mu\nu} = \partial_\mu v_\nu - \partial_\nu v_\mu - \frac{i}{2}[v_\mu, v_\nu],$$

and $\quad D_\mu\overline{\lambda} = \partial_\mu\overline{\lambda} - \frac{i}{2}[v_\mu, \overline{\lambda}].$

The reader should not confuse the gauge covariant derivative D_μ neither with the super covariant derivatives D_α and $\overline{D}_{\dot\alpha}$, nor with the auxiliary field D. The gauge group generators T^a satisfy,

$$[T^a, T^b] = i f^{abc} T^c$$

with real structure constants f^{abc}. The field strength then is $F^a_{\mu\nu} = \partial_\mu v^a_\nu - \partial_n v^a_\mu + \frac{1}{2} f^{abc} v^b_\mu v^c_\nu$.

We introduce the gauge coupling constant g by scaling the superfield V and hence all of its component fields as,

$$V \to 2gV \Leftrightarrow v_\mu \to 2gv_\mu,\ \lambda \to 2g\lambda, D \to 2gD$$

so that then we have the rescaled definitions of gauge covariant derivative and field strength,

$$D_\mu\lambda = \partial_\mu\lambda - ig[v_\mu, \lambda] \Rightarrow (D_\mu\lambda)^a = \partial_\mu\lambda^a + g f^{abc} v^b_\mu \lambda^c$$

$$F_{\mu\nu} = \partial_\mu v_\nu - \partial_\nu v_\mu - ig[v_\mu, v_\nu] \Rightarrow F^a_{\mu\nu} = \partial_\mu v^a_\nu - \partial_\nu v^a_\mu + g f^{abc} v^b_\mu v^c_\nu.$$

(We have implicitly assumed that the gauge group is simple so that there is a single coupling constant g. The generalisation to $G = G_1 \times G_2 \times \ldots$ and several $g_1, g_2, \ldots$ is straightforward.) Then the component expansion $W_\alpha = -i\lambda_\alpha(y) + \theta_\alpha D(y) + i(\sigma^{\mu\nu}\theta)_\alpha F_{\mu\nu}(y) + \theta\theta(\sigma^\mu\partial_\mu\overline{\lambda}(y))_\alpha$ of W_α remains unchanged, except for two things: there is on overall factor 2g multiplying the r.h.s. and $F_{\mu\nu}$ and $D_\mu\lambda$ are now given by $D_\mu\lambda = \partial_\mu\lambda - ig[v_\mu, \lambda] \Rightarrow (D_\mu\lambda)^a = \partial_\mu\lambda^a + g f^{abc} v^b_\mu \lambda^c$,

$F_{\mu\nu} = \partial_\mu v_\nu - \partial_\nu v_\mu - ig[v_\mu, v_\nu] \Rightarrow F^a_{\mu\nu} = \partial_\mu v^a_\nu - \partial_\nu v^a_\mu + g f^{abc} v^b_\mu v^c_\nu$. It follows that $\int d^2\theta\, W^\alpha W_\alpha = -\frac{1}{2} f_{\mu\nu} f^{\mu\nu} - 2i\lambda\sigma^\mu \partial_\mu \bar{\lambda} + D^2 + \frac{i}{4} \epsilon^{\mu\nu\rho\sigma} f_{\mu\nu} f_{\rho\sigma}$ also remains unchanged except for the replacements $f_{\mu\nu} \to F_{\mu\nu}$ and $\partial_\mu \bar{\lambda} \to D_\mu \bar{\lambda}$ and an overall factor $4g^2$. One then introduces the complex coupling constant, $\tau = \frac{\Theta}{2\pi} + \frac{4\pi i}{g^2}$ where Θ stands for the Θ-angle. (We use a capital Θ to avoid confusion with the superspace coordinates θ.) Then,

$$\mathcal{L}_{\text{gauge}} = \frac{1}{32} \pi \,\text{Im}\left(\tau \int d^2\theta\, \text{Tr}\, W^\alpha W_\alpha\right)$$

$$= \text{Tr}\left(-\frac{1}{4} F_{\mu\nu} F^{\mu\nu} - i\lambda\sigma^\mu D_\mu \bar{\lambda} + \frac{1}{2} D^2\right) + \frac{\Theta}{32\pi^2} g^2\, \text{Tr}\, F_{\mu\nu} \tilde{F}^{\mu\nu}$$

where, $\tilde{F}^{\mu\nu} = \frac{1}{2} \epsilon^{\mu\nu\rho\sigma} F_{\rho\sigma}$ is the dual field strength. The single term $\text{Tr}\, W^\alpha W_\alpha$ has produced both, the conventionally normalised gauge kinetic term $-\frac{1}{4} \text{Tr}\, F_{\mu\nu} F^{\mu\nu}$ and the instanton density $\frac{g^2}{32\pi^2} \text{Tr}\, F_{\mu\nu} \tilde{F}^{\mu\nu}$ which multiplies the Θ-angle.

N = 1 Gauge Theory with Matter

We now add chiral (matter) multiplets ϕ^i transforming in some representation R of the gauge group where the generators are represented by matrices $\left(T^a_R\right)^i_j$. Then, $\phi^i \to \left(e^{i\Lambda}\right)^i_j \phi^j$, $\phi^\dagger_i \to \phi^\dagger_j \left(e^{-i\Lambda\dagger}\right)^j_i$ or simply $\phi \to e^{i\Lambda}\phi$, $\phi^\dagger \to \phi^\dagger e^{-i\Lambda\dagger}$ where $\Lambda = \Lambda^a T^a_R$ is understood. Then, $\phi^\dagger e^V \phi \equiv \phi^\dagger e^{Va} T^a_R \phi \equiv \phi^\dagger_i \left(e^V\right)^i_j \phi^j$ is the gauge invariant generalisation of the kinetic term and, $\mathcal{L}_{\text{matter}} = \int d^2\theta d^2\bar{\theta}\, \phi^\dagger e^V \phi + \int d^2\theta\, W(\phi) + \int d^2\bar{\theta} \left[W(\phi)\right]^\dagger$.

We have not yet scaled V by 2g, or equivalently we set 2g = 1 for the time being to simplify the formula. We want to compute the $\theta\theta\bar{\theta}\bar{\theta}$ component (D-term) of $\phi^\dagger e^V \phi = \phi^\dagger \phi + \phi^\dagger V \phi + \frac{1}{2} \phi^\dagger V^2 \phi$. The first term is given by $\phi^\dagger \phi\big|_{\theta\theta\bar{\theta}\bar{\theta}} = -\frac{1}{4} z^\dagger d^2 z - \frac{1}{4} d^2 z^\dagger z + \frac{1}{2} \partial_\mu z^\dagger \partial^\mu z + f^\dagger f + \frac{i}{2} \partial_\mu \psi\sigma^\mu \bar{\psi} - \frac{i}{2} \psi\sigma^\mu \partial_\mu \bar{\psi}$ $= \partial_\mu z^\dagger \partial^\mu z + \frac{i}{2} \left(\partial_\mu \psi\sigma^\mu \bar{\psi} - \psi\sigma^\mu \partial_\mu \bar{\psi}\right) + f^\dagger f + \text{total derivative}$. The second term is, $\phi^\dagger V \phi\big|_{\theta\theta\bar{\theta}\bar{\theta}} = \frac{i}{2} z^\dagger v^\mu \partial_\mu z - \frac{i}{2} \partial_\mu z^\dagger v^\mu z - \frac{1}{2} \bar{\psi} \bar{\sigma}^\mu v_\mu \psi + \frac{i}{\sqrt{2}} z^\dagger \lambda \psi - \frac{i}{\sqrt{2}} \bar{\psi} \bar{\lambda} z + \frac{1}{2} z^\dagger D z$ and the third term is,

$$\phi^\dagger V^2 \phi\big|_{\theta\theta\bar{\theta}\bar{\theta}} = \frac{1}{4} z^\dagger v^\mu v_\mu z.$$

Combining all three terms gives,

$$\phi^{\dagger}e^{V}\phi\big|_{\theta\theta\bar{\theta}\bar{\theta}} = \left(D_{\mu}z\right)^{\dagger}D^{\mu}z - i\bar{\psi}\bar{\sigma}^{\mu}D_{\mu}\psi + f^{\dagger}f$$
$$+\frac{i}{\sqrt{2}}z^{\dagger}\lambda\psi - \frac{i}{\sqrt{2}}\bar{\psi}\bar{\lambda}z + \frac{1}{2}z^{\dagger}Dz + \text{total derivative.}$$

With $D_{\mu}z = \partial_{\mu}z - \frac{i}{2}v_{\mu}^{a}T_{R}^{a}z$ and $D_{\mu}\psi = \partial_{\mu}\mu - \frac{i}{2}v_{\mu}^{a}T_{R}^{a}\psi$ .Wenowrescale $V \to 2gV$ andusethe firstidentity $\chi\sigma^{\mu}\bar{\psi} = -\bar{\psi}\bar{\sigma}^{\mu}\chi$, $\chi\sigma^{\mu}\bar{\sigma}^{\nu}\psi = \psi\sigma^{\nu}\bar{\sigma}^{\mu}\chi$, $\left(\chi\sigma^{\mu}\bar{\psi}\right)^{\dagger} = \psi\sigma^{\mu}\bar{\chi}$,$\left(\chi\sigma^{\mu}\bar{\sigma}^{\nu}\psi\right)^{\dagger} = \bar{\psi}\bar{\sigma}^{\nu}\sigma^{\mu}\bar{\chi}$, $\psi\chi = \chi\psi$, $\bar{\psi}\bar{\chi} = \bar{\chi}\bar{\psi}$, $(\psi\chi)^{\dagger} = \bar{\psi}\bar{\chi}$ to rewrite $\bar{\psi}\bar{\sigma}^{\mu}D_{\mu}\psi = \psi\sigma^{\mu}D_{\mu}\bar{\psi}$ + total derivative. Then this is replaced by,

$$\phi^{\dagger}e^{2gV}\phi\big|_{\theta\theta\bar{\theta}\bar{\theta}} = \left(D_{\mu}z\right)^{\dagger}D^{\mu}z - i\psi\sigma^{\mu}D_{\mu}\bar{\psi} + f^{\dagger}f$$
$$+i\sqrt{2}\,gz^{\dagger}\lambda\psi - i\sqrt{2}\,g\bar{\psi}\bar{\lambda}z + gz^{\dagger}Dz + \text{total derivative.}$$

Now with $D_{\mu}z = \partial_{\mu}z - ig\,v_{\mu}^{a}T_{R}^{a}z$ and $D_{\mu}\psi = \partial_{\mu}\psi - ig\,v_{\mu}^{a}T_{R}^{a}\psi$. This part of the Lagrangian contains the kinetic terms for the scalar fields and the matter fermions ψ^{i}, as well as specific interactions between the z^{i}, the ψ^{i} and the gauginos λ^{a}. One has e.g. $z^{\dagger}\lambda\psi \equiv z_{i}^{\dagger}\left(T_{R}^{a}\right)_{j}^{i}\lambda^{a}\psi^{j}$.

What happens to the superpotential $W(\phi)$? This must be a chiral superfield and hence must be constructed from the ϕ^{i} alone. It must also be gauge invariant which imposes severe constraints on the superpotential. A term of the form $a_{i_1,\ldots i_n}\phi^{i_1}\ldots\phi^{i_n}$ will only be allowed if the n-fold product of the representation R contains the trivial representation and then $a_{i_1,\ldots i_n}$ must be an invariant tensor of the gauge group. An example is G = SU(3) with R = 3. Then 3×3×3 = 1+. . . and the correponding SU(3) invariant tensor is ϵ_{ijk}. In this example, however, bilinears would not be gauge invariant. On the other hand, the representation R need not be irreducible. Taking again the example of G = SU(3) one may have $R = 3\oplus\bar{3}$ corresponding to a chiral superfield ϕ^{i} transforming as 3 ("quark") and a chiral superfield $\tilde{\phi}^{i}$ transforming as $\bar{3}$ ("antiquark"). Then one can form the gauge invariant chiral superfield $\tilde{\phi}^{i}\phi^{i}$ which corresponds to a "quark" mass term.

There is a last type of term that may appear in case the gauge group simply is U(1) or contains U(1) factors. If there is at least an extra U(1) factor the gauge group certainly is not simple and we have several coupling constants: These are the Fayet-Iliopoulos terms. Let V^{A} denote the vector superfield in the abelian case, or the component corresponding to an abelian factor. Then under an abelian gauge transformation, $V^{A} \to V^{A} - i\Lambda + i\Lambda^{\dagger}$, with Λ

a chiral superfield. From the component expansion of such a chiral or anti chiral superfield $\phi(y,\theta)=z(x)+\sqrt{2}\theta\psi(x)+i\theta\sigma^{\mu}\bar{\theta}\partial_{\mu}z(x)-\theta\theta f(x)-\frac{i}{\sqrt{2}}\theta\theta\partial_{\mu}\psi(x)\sigma^{\mu}\bar{\theta}-\frac{1}{4}\theta\theta\bar{\theta}\bar{\theta}\partial^{2}z(x)$ or $\phi(y,\bar{\theta})=\bar{z}(\bar{y})+\sqrt{2}\bar{\theta}\bar{\psi}(\bar{y})-\bar{\theta}\bar{\theta}\bar{f}(\bar{y})\ldots+\frac{i}{\sqrt{2}}\bar{\theta}\bar{\theta}\theta\sigma^{\mu}\partial_{\mu}\bar{\psi}(x)-\frac{1}{4}\theta\theta\bar{\theta}\bar{\theta}\partial^{2}\bar{z}(x)$ one sees that the D-term (the term $\sim\theta\theta\bar{\theta}\bar{\theta}$) transforms as $D^{A}\to D^{A}+\partial_{\mu}\partial^{\mu}(\ldots)$, i.e. as a total derivative. Being a D-term, it also transforms as a total derivative under susy. It follows that,

$$\mathcal{L}_{FI}=\sum_{A\in\text{abelian factors}}\xi^{A}\int d^{2}\theta d^{2}\bar{\theta}V^{A}=\frac{1}{2}\sum_{A\in\text{abelian factors}}\xi^{A}D^{A}$$

is a susy and gauge invariant Lagrangian (i.e. up to total derivatives). We can finally write the full N = 1 Lagrangian, being the sum of

$$\mathcal{L}_{gauge}=\frac{1}{32}\pi\,\mathrm{Im}\left(\tau\int d^{2}\theta\,\mathrm{Tr}\,W^{\alpha}W_{\alpha}\right)=\mathrm{Tr}\left(-\frac{1}{4}F_{\mu\nu}F^{\mu\nu}-i\lambda\sigma^{\mu}D_{\mu}\bar{\lambda}+\frac{1}{2}D^{2}\right)+\frac{\Theta}{32\pi^{2}}g^{2}\,\mathrm{Tr}\,F_{\mu\nu}\tilde{F}^{\mu\nu},$$

$$\phi^{\dagger}e^{2gV}\phi\big|_{\theta\theta\bar{\theta}\bar{\theta}}=(D_{\mu}z)^{\dagger}D^{\mu}z-i\psi\sigma^{\mu}D_{\mu}\bar{\psi}+f^{\dagger}f+i\sqrt{2}gz^{\dagger}\lambda\psi-i\sqrt{2}g\bar{\psi}\bar{\lambda}z+gz^{\dagger}Dz+\text{total derivative}$$

and $\mathcal{L}_{FI}=\sum_{A\in\text{abelian factors}}\xi^{A}\int d^{2}\theta d^{2}\bar{\theta}V^{A}=\frac{1}{2}\sum_{A\in\text{abelian factors}}\xi^{A}D^{A}$:

$$\begin{aligned}\mathcal{L}=\mathcal{L}_{gauge}+\mathcal{L}_{matter}+\mathcal{L}_{FI}&=\frac{1}{32\pi}\mathrm{Im}\left(\tau\int d^{2}\theta\,\mathrm{Tr}\,W^{\alpha}W_{\alpha}\right)+2g\sum\nolimits_{A}\xi^{A}\int d^{2}\theta d^{2}\bar{\theta}V^{A}\\&+\int d^{2}\theta d^{2}\bar{\theta}\phi^{\dagger}e^{2gV}\phi+\int d^{2}\theta W(\phi)+\int d^{2}\bar{\theta}\left[W(\phi)\right]^{\dagger}\\&=\mathrm{Tr}\left(-\frac{1}{4}F_{\mu\nu}F^{\mu\nu}-i\lambda\sigma^{\mu}D_{\mu}\bar{\lambda}+\frac{1}{2}D^{2}\right)+\frac{\Theta}{32\pi^{2}}g^{2}\mathrm{Tr}\,F_{\mu\nu}\tilde{F}^{\mu\nu}+g\sum\nolimits_{A}\xi^{A}D^{A}\\&+(D_{\mu}z)^{\dagger}D^{\mu}z-i\psi\sigma^{\mu}D_{\mu}\psi+f^{\dagger}f+i\sqrt{2}gz^{\dagger}\lambda\psi-i\sqrt{2}g\bar{\psi}\bar{\lambda}z+gz^{\dagger}Dz\\&-\frac{\partial W}{\partial z^{i}}f^{i}+\text{h.c.}-\frac{1}{2}\frac{\partial^{2}W}{\partial z^{i}\partial z^{j}}\psi^{i}\psi^{j}+\text{h.c.}+\text{total derivative.}\end{aligned}$$

The auxiliary field equations of motion are, $f_{i}^{\dagger}=\frac{\partial W}{\partial z^{i}}$ and $D^{a}=-gz^{\dagger}T^{a}z-g\xi^{a}$ where it is understood that $\xi^{a}=0$ if a does not take values in an abelian factor of the gauge group. Substituting this back into the Lagrangian one finds,

$$\begin{aligned}\mathcal{L}=\mathrm{Tr}\left(-\frac{1}{4}F_{\mu\nu}F^{\mu\nu}-i\lambda\sigma^{\mu}D_{\mu}\bar{\lambda}\right)&+\frac{\Theta}{32\pi^{2}}g^{2}\mathrm{Tr}\,F_{\mu\nu}\tilde{F}^{\mu\nu}+(D_{\mu}z)^{\dagger}D^{\mu}z-i\psi\sigma^{\mu}D_{\mu}\bar{\psi}\\&+i\sqrt{2}gz^{\dagger}\lambda\psi-i\sqrt{2}g\bar{\psi}\bar{\lambda}z-\frac{1}{2}\frac{\partial^{2}W}{\partial z^{i}\partial z^{j}}\psi^{i}\psi^{j}-\frac{1}{2}\left(\frac{\partial^{2}W}{\partial z^{i}\partial z^{j}}\right)^{\dagger}\bar{\psi}^{i}\bar{\psi}^{j}\\&-V(z^{\dagger},z)+\text{total derivative,}\end{aligned}$$

where the scalar potential $V(z^\dagger, z)$ is given by,

$$V(z^\dagger, z) = f^\dagger f + \frac{1}{2}D^2 = \sum_i \left|\frac{\partial W}{\partial z^i}\right|^2 + \frac{g^2}{2}\sum_a \left|z^\dagger T^a z + \xi^a\right|^2.$$

Supersymmetric QCD

At this point we have all the ingredients to write the action for supersymmetric QCD. The gauge group is SU(3). (More generally, one could consider SU(N).) There are then gauge bosons v_μ^a, $a = 1, \dots 8$ called the gluons, as well as their supersymmetric partners, the 8 gauginos or gluinos λ^a. If one considers pure N = 1 "glue", this is all there is. To describe N = 1 QCD however, one also has to add quarks transforming in the 3 of SU(3) as well as antiquarks in the $\bar{3}$. They are associated with chiral superfields. More precisely there are chiral superfields $Q_L^i = q_L^i + \sqrt{2}\theta\psi_L^i - \theta\theta\, f_L^i$, $i = 1, 2, 3$, and $L = 1, \dots N_f$ labels the flavours. These fields transform in the 3 representation of the gauge group and correspond to left-handed quarks (or right-handed antiquarks). There are also chiral superfields $\tilde{Q}_{i,L} = \tilde{q}_{iL} + \sqrt{2}\theta\,\tilde{\psi}_{iL} - \theta\theta\,\tilde{f}_{iL}$, $i = 1,2,3$ and $L = 1, \dots N_f$. They transform in the $\bar{3}$ representation of the gauge group and correspond to left-handed antiquarks (or right-handed quarks).

The gauge group does not contain any U(1) factor, and hence no Fayet-Iliopoulos term can appear. The component Lagrangian for massless susy QCD then is given by $\mathcal{L} = \mathrm{Tr}\left(-\frac{1}{4}F_{\mu\nu}F^{\mu\nu} - i\lambda\sigma^\mu D_\mu\bar{\lambda}\right)\dots - \frac{1}{2}\left(\frac{\partial^2 W}{\partial z^i \partial z^j}\right)^\dagger \bar{\psi}^i\bar{\psi}^j - V(z^\dagger, z) + \text{total derivative}$, with $z = (q, \tilde{q})$, $\xi^a = 0$ and vanishing superpotential. Since all terms in the Lagrangian are diagonal in the flavour indices of the quarks and separately in the flavour indices of the antiquarks, there is an $SU(N_f)_L \times SU(N_f)_R$ global symmetry. In addition there is a $U(1)_V$ acting as $Q \to e^{iv}Q$, $\tilde{Q} \to e^{-iv}\tilde{Q}$, as well as an $U(1)_A$ acting as $Q \to e^{ia}Q$, $\tilde{Q} \to e^{-ia}\tilde{Q}$. We also have the global $U(1)_R$ symmetry acting as $Q(x,\theta) \to e^{-iq}Q(x, e^{iq}\theta)$, $\tilde{Q}(x,\theta) \to e^{-iq}\tilde{Q}(x, e^{iq}\theta)$ and $V(x,\theta,\bar{\theta}) \to V(x, e^{iq}\theta, e^{-iq}\bar{\theta})$. Thus the global symmetry group in the massless case is $U(N_f)_L \times U(N_f)_R \times U(1)_R$.

Due to the presence of both representations 3 and $\bar{3}$ of the gauge group, one may add a gauge invariant superpotential,

$$W(Q, \tilde{Q}) = m_{L,M} Q_L^i \tilde{Q}_{i,M}.$$

This is a quark mass term and $m_{L,M}$ is the $N_f \times N_f$ mass matrix. Using the global symmetry of the other terms in the Lagrangian (which is just the massless Lagrangian) one can diagonalise the superpotential so that it reads,

$$W(Q,\tilde{Q}) = \sum_L m_L Q_L^i \tilde{Q}_{i,L}.$$

For the gauge group SU(3) one could also add the gauge invariant terms $a_{LMN} \epsilon_{ijk} Q_L^i Q_M^j Q_N^k$ and $\tilde{a}_{LMN} \epsilon_{ijk} \tilde{Q}_{iL} \tilde{Q}_{jM} \tilde{Q}_{kN}$. However, they explicitly violate baryon number conservation and will not be considered. Then finally one arrives at the following Lagrangian (where we suppress as much as possible all gauge and flavour indices):

$$\begin{aligned}\mathcal{L} = {} & \mathrm{Tr}\left(-\frac{1}{4}F_{\mu\nu}F^{\mu\nu} - i\lambda\sigma^{\mu}D_{\mu}\bar{\lambda}\right) + \frac{\Theta}{32\pi^2}g^2 \mathrm{Tr}\,F_{\mu\nu}\tilde{F}^{\mu\nu} \\ & + (D_{\mu}q)^{\dagger} D^{\mu}q + (D_{\mu}\tilde{q})^{\dagger} D^{\mu}\tilde{q} - i\psi\sigma^{\mu}D_{\mu}\bar{\psi} - i\tilde{\psi}\sigma^{\mu}D_{\mu}\bar{\tilde{\psi}} \\ & + i\sqrt{2}gq^{\dagger}\lambda\psi + i\sqrt{2}g\tilde{q}^{\dagger}\lambda\tilde{\psi} - i\sqrt{2}g\overline{\psi\lambda}q - i\sqrt{2}g\overline{\tilde{\psi}\lambda}\tilde{q} \\ & - \frac{1}{2}\sum\nolimits_L m_L\left(\psi_L\psi_L + \bar{\tilde{\psi}}_L\bar{\tilde{\psi}}_L\right) - V(q,\tilde{q},q^{\dagger},\tilde{q}^{\dagger}) + \text{total derivative},\end{aligned}$$

where the scalar potential V is given by,

$$V(q,\tilde{q},q^{\dagger},\tilde{q}^{\dagger}) = \sum_{L=1}^{N_f} m_L^2\left(q_L^{\dagger}q_L + \tilde{q}_L^{\dagger}\tilde{q}_L\right) + \frac{g^2}{2}\sum_{a=1}^{8}\left|q^{\dagger}T^a q + \tilde{q}^{\dagger}T^a\tilde{q}\right|^2.$$

References

- Gauge-theories: scholarpedia.org, Retrieved 23, June 2020
- Higgs-boson: britannica.com, Retrieved 02, January 2020
- Relativistic-quantum-field-theory-i-spring: mit.edu, Retrieved 19, May 2020
- String-theory: britannica.com, Retrieved 10, August 2020
- String-Theory-a-mini-course-231041410: researchgate.net Retrieved 27, April 2020

3

Subatomic Particles

The particles which are smaller than atoms are known as subatomic particles. These can be either elementary particles or composite particles. The subatomic particles which have an integer spin are known as bosons and those that have an off half-integer spin are known as fermions. This chapter discusses in detail these two types of subatomic particles.

The particles that have been identified in high-energy experiments fall into distinct classes. There are the leptons, all of which have spin 1/2. They may be charged or neutral. The charged leptons have electromagnetic as well as weak interactions; the neutral ones only interact weakly. There are three well-defined lepton pairs, the electron (e^-) and the electron neutrino (νe), the muon (μ^-) and the muon neutrino (ν_μ), and the (much heavier) charged lepton, the tau (τ), and its tau neutrino (ν_τ). These particles all have antiparticles, in accordance with the predictions of relativistic quantum mechanics. There appear to exist approximate "lepton-type" conservation laws: the number of e^- plus the number of ν_e minus the number of the corresponding antiparticles e + and $\bar{\nu}_e$ is conserved in weak reactions, and similarly for the muon and tau-type leptons. These conservation laws would follow automatically in the standard model if the neutrinos are massless. Recently, however, evidence for tiny nonzero neutrino masses and subtle violation of these conservations laws has been observed. There is no understanding of the hierarchy of masses in Table below or why the observed neutrinos are so light.

In table, the leptons, Charges are in units of the positron (e^+) charge $e = 1.602 \times 10^{-19}$ coulomb. In addition to the upper limits, two of the neutrinos have masses larger than 0.05 eV/c^2 and 0.005 eV/c^2, respectively. The ν_e, ν_μ, and ν_τ are mixtures of the states of definite mass.

Particle	Q	Mass
e^-	-1	0.51 MeV/c^2
μ	-1	0105.7 MeV/c^2
τ^-	-1	1777 MeV/c^2
ν_e	0	<0.15 eV/c^2
ν_μ	0	<0.15 eV/c^2
ν_τ	0	<0.15 eV/c^2

In addition to the leptons there exist hadrons, which have strong interactions as well as the electromagnetic and weak. These particles have a variety of spins, both integral and half-integral, and their masses range from the value of 135 MeV/c^2 for the neutral pion π^o to 11 020 MeV/c² for one of the upsilon (heavy quark) states. The particles with half-integral spin are called baryons, and there is clear evidence for baryon conservation: The number of baryons minus the number of antibaryons is constant in any interaction. The best evidence for this is the stability of the lightest baryon, the proton (if the proton decays, it does so with a lifetime in excess of 10^{33} yr). In contrast to charge conservation, there is no deep principle that makes baryon conservation compelling, and it may turn out that baryon conservation is only approximate. The particles with integer spin are called mesons, and they have baryon number B = 0. There are hundreds of different kinds of hadrons, some almost stable and some (known as resonances) extremely short-lived. The degree of stability depends mainly on the mass of the hadron. If its mass lies above the threshold for an allowed decay channel, it will decay rapidly; if it does not, the decay will proceed through a channel that may have a strongly suppressed rate, e. g., because it can only be driven by the weak or electromagnetic interactions.

The large number of hadrons has led to the universal acceptance of the notion that the hadrons, in contrast to the leptons, are composite. In particular, experiments involving lepton–hadron scattering or $e^+ e^-$ annihilation into hadrons have established that hadrons are bound states of point-like spin- 1 2 particles of fractional charge, known as quarks. Six types of quarks have been identified. As with the leptons, there is no understanding of the extreme hierarchy of quark masses. For each type of quark there is a corresponding antiquark. Baryons are bound states of three quarks (e. g., proton = uud; neutron = udd), while mesons consist of a quark and an antiquark. Matter and decay processes under normal terrestrial conditions involve only the e^-, ν_e, u, and d. These four types of fundamental particle are replicated in two heavier families, (μ^-, ν_μ, c, s) and (τ^-, ν_τ, t, b). The reason for the existence of these heavier copies is still unclear.

Table: The quarks (spin-$\frac{1}{2}$ constituents of hadrons). Each quark carries baryon number B $=\frac{1}{3}$, while the antiquarks have B $= -\frac{1}{3}$.

Particle	Q	Mass
u (up)	$\frac{2}{3}$	1.5-5 MeV/c²
d (down)	$-\frac{1}{3}$	5-9 MeV/c²
s (strange)	$-\frac{1}{3}$	80-155 MeV/c²
c (charm)	$\frac{2}{3}$	1-1.14 GeV/c²

b (bottom)	$-\frac{1}{3}$	4-4.5 GeV/c^2
b (bottom)	$\frac{2}{3}$	175-180 Ge V/c^2

Classification of Interactions

The interactions fall into four types, the electromagnetic, weak, and strong, and the gravitational interaction. If we take the proton mass as a standard, the last is 10^{-36} times the strength of the electromagnetic interaction, and will mainly be neglected in what follows. (The unification of gravity with the other interactions is one of the major outstanding goals.) The first two interactions were most cleanly explored with the leptons, which do not have strong interactions that mask them.

Electromagnetic Interactions

The electromagnetic interactions of charged leptons (electron, muon, and tau) are best described in terms of equations of motion, derived from a Lagrangian function, which are solved in a power series in the fine-structure constant $e^2/4\pi\hbar c = \alpha \simeq 1/137$, a small parameter. The Lagrangian density consists of a term that describes the free-photon field,

$$\mathcal{L}_\gamma = -\frac{1}{4}F_{\mu\nu}(x)F^{\mu\nu}(x)$$

Where,

$$F_{\mu\nu}(x) = \frac{\partial A_\nu(x)}{\partial x^\mu} - \frac{\partial A_\mu(x)}{\partial x^\nu}$$

is the electromagnetic field tensor. $\mathcal{L}_\gamma$ is just $\frac{1}{2}\left[E^2(x) - B^2(x)\right]$ in more common notation. It is written in terms of the vector potential $A_\mu(x)$ because the terms that involve the lepton and its interaction with the electromagnetic field are simplest when written in terms of $A_\mu(x)$:

$$\mathcal{L}_1 = i\bar{\psi}(x)\gamma^\alpha\left(\frac{\partial}{\partial x^\alpha} - ieA_\alpha(x)\right)\psi(x) - m\bar{\psi}(x)\psi(x)$$

Here $\psi(x)$ is a four-component spinor representing the electron, muon, or tau, $\bar{\psi}(x)$ $\bar{\psi}(x) = \bar{\psi}^\dagger(x)\gamma^0$, the $\gamma^\alpha(\alpha = 0, 1, 2, 3)$ are the Dirac matrices [4 × 4 matrices that satisfy the conditions $(\gamma^1)^2 = (\gamma^2)^2 = (\gamma^3)^2 = -(\gamma^0)^2 = -1$ and $\gamma^\alpha\gamma^\beta = -\gamma^\beta\gamma^\alpha$ for $\beta \neq \alpha$]; m has the dimensions of a mass in the natural units in which ~ = c = 1. If e were zero, the Lagrangian would describe a free lepton; with e 6= o the interaction has the form:

$$-eA^{\alpha}(x)j_{\alpha}(x)$$

where the current $j_{\alpha}(x)$ is given by,

$$j_{\alpha}(x)=-\bar{\psi}(x)\gamma_{\alpha}\psi(x)$$

The equations of motion show that the current is conserved,

$$\frac{\partial}{\partial x_{\alpha}}j_{\alpha}(x)=0$$

so that the charge:

$$Q=\int d^3r\, j_0(r,t)$$

is a constant of the motion.

The form of the interaction is obtained by making the replacement:

$$\frac{\partial}{\partial x^{\alpha}}\rightarrow\frac{\partial}{\partial x^{\alpha}}-ieA_{\alpha}(x)$$

in the Lagrangian for a free lepton. This minimal coupling follows from a deep principle, local gauge invariance. The requirement that ψ(x) can have its phase changed locally without affecting the physics of the lepton, that is, invariance under:

$$\psi(x)\rightarrow e^{-i\theta(x)}\psi(x)$$

can only be implemented through the introduction of a vector field $A_{\alpha}(x)$, coupled as in $\frac{\partial}{\partial x^{\alpha}}\rightarrow\frac{\partial}{\partial x^{\alpha}}-ieA_{\alpha}(x)\frac{\partial}{\partial x^{\alpha}}\rightarrow\frac{\partial}{\partial x^{\alpha}}-ieA_{\alpha}(x)$, and transforming according to:

$$A_{\alpha}(x)\rightarrow A_{\alpha}(x)-\frac{1}{e}\frac{\partial\theta(x)}{\partial x^{\alpha}}$$

This dictates that the free-photon Lagrangian density contains only the gauge invariant combination $F_{\mu\nu}(x)=\frac{\partial A_{\nu}(x)}{\partial x^{\mu}}-\frac{\partial A_{\mu}(x)}{\partial x^{\nu}}$, and that terms of the form $M^2A_{\alpha}^2(x)$ be absent. Thus local gauge invariance is a very powerful requirement; it implies the existence of a massless vector particle (the photon, γ), which mediates a long-range force. It also fixes the form of the coupling and leads to charge conservation, and implies masslessness of the photon. The resulting is in extremely good agreement with experiment, as

Table shows. In working out the consequences of the equations of motion that follow from $\mathcal{L}_1 = i\bar{\psi}(x)\gamma^{\alpha}\left(\frac{\partial}{\partial x^{\alpha}} - ieA_{\alpha}(x)\right)\psi(x) - m\bar{\psi}(x)\psi(x)$, infinities appear, and the theory seems not to make sense. The work of S. Tomonaga, J. Schwinger, R. P. Feynman, and F. J. Dyson in the late 1940s clarified the nature of the problem and showed a way of eliminating the difficulties. In creating renormalization theory these authors pointed out that the parameters e and m that appear in $\mathcal{L}_1 = i\bar{\psi}(x)\gamma^{\alpha}\left(\frac{\partial}{\partial x^{\alpha}} - ieA_{\alpha}(x)\right)\psi(x) - m\bar{\psi}(x)\psi(x)$ can be identified as the charge and the mass of the lepton only in lowest order. When the charge and mass are calculated in higher order, infinite integrals appear. After a rescaling of the lepton fields, it turns out that these are the only infinite integrals in the theory. Thus by absorbing them into the definitions of new quantities, the renormalized (i. e., physically measured) charge and mass, all infinities are removed, and the rest of the theoretically calculated quantities are finite. Gauge invariance ensures that in the renormalized theory the current is still conserved, and the photon remains massless (the experimental upper limit on the photon mass is 6×10^{-17} eV/c²).

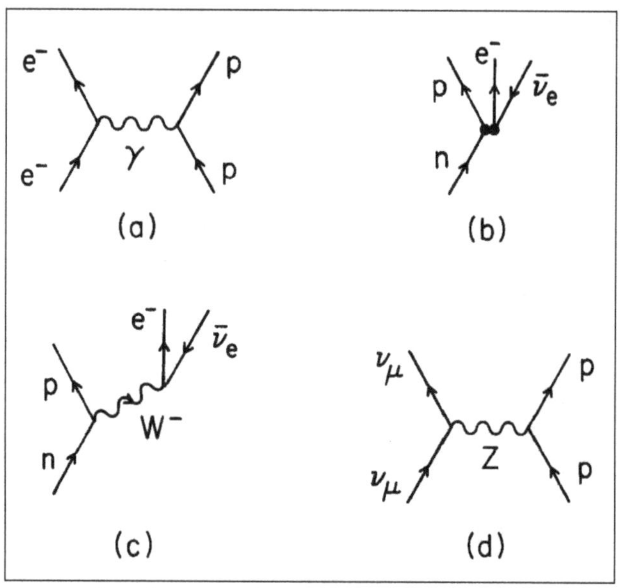

(a) Long-range force between electron and proton mediated by a photon.
(b) Four-fermi (zero-range) description of beta decay ($n \rightarrow pe^- \bar{V}_e$).
(c) Beta decay mediated by a W⁻. (d) A neutral current process mediated by the Z.

In Table, Extraction of the (inverse) fine structure constant α^{-1} from various experiments. The consistency of the various determinations tests QED. The numbers in parentheses (square brackets) represent the uncertainty in the last digits (the fractional uncertainty). The last column is the difference from the (most precise) value $\alpha^{-1}(a_e)$ in the first row. A precise measurement of the muon gyromagnetic ratio a_μ is ~ 2.4σ above the theoretical prediction, but that quantity is more sensitive to new (TeV-scale) physics.

Experiment	Value of α^{-1}		Difference from α^{-1} (a_e)
Deviation from gyromagnetic ratio, $a_e = (g - 2)/2$ for e^-	137.035 999 58 (52)	$[3.8 \times 10^{-9}]$	-
ac Josephson effect	137.035 988 0 (51)	$[3.7 \times 10^{-8}]$	$(0.116 \pm 0.051) \times 10^{-4}$
h/m_n (m_n is the neutron mass)	137.036 011 9 (51)	$[3.7 \times 10^{-8}]$	$(-0.123 \pm 0.051) \times 10^{-4}$
Hyperfine structure in muonium, μ^+e^-	137.035 993 2 (83)	$[6.0 \times 10^{-8}]$	$(0.064 \pm 0.083) \times 10^{-4}$
Cesium D_1 line	137.035 992 4 (41)	$[3.0 \times 10_{-8}]$	$(0.072 \pm 0.041) \times 10^{-4}$

The possibility of absorbing the divergences of a theory in a finite number of renormalizations of physical quantities is limited to a small class of theories, e. g., those involving the coupling of spin- 1/2 to spin-0 particles with a very restrictive form of the coupling. Theories involving vector (spin-1) fields are only renormalizable when the couplings are minimal and local gauge invariance holds. Thus gauge-invariant couplings like $\bar{\psi}(x)\gamma^\alpha\gamma^\beta\psi(x)F_{\alpha\beta}(x)$, which are known not to be needed in quantum electrodynamics, are eliminated by the requirement of renormalizability. (The apparent infinities for non-renormalizable theories become finite when the theories are viewed as a low energy approximation to a more fundamental theory. In that case, however, the low energy predictions have a very large sensitivity to the energy scale at which the new physics appears.) The electrodynamics of hadrons involves a coupling of the form:

$$-eA^\alpha(x)j_\alpha^{had}(x)$$

For one-photon processes, such as photoproduction (e. g., $\gamma p \rightarrow \pi^o p$), matrix elements of the conserved current $j_\alpha^{had}(x)$ are measured to first order in e, while for two-photon processes, such as hadronic Compton scattering ($\gamma p \rightarrow \gamma p$), matrix elements of products like $j_\alpha^{had}(x)\ j_\beta^{had}(y)$ enter. Within the quark theory one can write an explicit form for the hadronic current:

$$j_\alpha^{had}(x) = \frac{2}{3}\bar{u}\gamma^\alpha u - \frac{1}{3}\bar{d}\gamma^\alpha d - \frac{1}{3}\bar{s}\gamma^\alpha s...$$

where we use particle labels for the spinor operators (which are evaluated at x), and the coefficients are just the charges in units of e. The total electromagnetic interaction is therefore $-eA^\alpha j_\alpha^\gamma$, where:

$$j_\alpha^\gamma = j_\alpha + j_\alpha^{had} = \sum_i Q_i \bar{\psi}_i \gamma_\alpha \psi_i,$$

and the sum extends over all the leptons and quarks (ψ_i = e, μ, τ, ν_e, ν_μ, ν_τ, u, d, c, s, b, t), and where Q_i is the charge of ψ_i.

Weak Interactions

In contrast to the electromagnetic interaction, it took many decades of experimental and theoretical work to arrive at a compact phenomenological Lagrangian density describing the weak interactions. The form:

$$\mathcal{L}_W = -\frac{G}{\sqrt{2}} J^{\dagger}_{\alpha}(x) J^{\alpha}(x)$$

involves vectorial quantities, as originally proposed by E. Fermi. The current $J^{\alpha}(x)$ is known as a charged current since it changes (lowers) the electric charge when it acts on a state. That is, it describes a transition such as $\nu_e \rightarrow e^-$ of one particle into another, or the corresponding creation of an $e^-\bar{\nu}_e$ pair. Similarly, $J^{\dagger}_{\alpha}$ describes a charge-raising transition such as $n \rightarrow p$. Equation above describes a zero-range four-fermi interaction, in contrast to electrodynamics, in which the force is transmitted by the exchange of a photon. An additional class of "neutral-current" terms was discovered in 1973. $J^{\alpha}(x)$ consists of leptonic and hadronic parts:

$$J^{\alpha}(x) = J^{\alpha}_{\text{lept}}(x) + j^{\alpha}_{\text{had}}(x)$$

Thus, it describes purely leptonic interactions, such as:

$$\mu^- \rightarrow e^- + \bar{\nu}_e + \nu_\mu$$

$$\nu_\mu + e^- \rightarrow \nu_e + \mu^-$$

through terms quadratic in J_{lept}; semileptonic interactions, most exhaustively studied in decay processes such as:

$$n \rightarrow p + e^- + \bar{\nu}_e \,(\text{beta decay})$$

$$\pi^+ \rightarrow \mu^+ + \nu_\mu$$

$$\Lambda^o \rightarrow p + e^- + \bar{\nu}_e$$

and more recently in neutrino-scattering reactions such as:

$$\nu_\mu + n \rightarrow \mu^- + p\,(\text{or } \mu^- + \text{hadrons})$$

$$\bar{\nu}_\mu + p \rightarrow \mu^+ + n\,(\text{or } \mu^+ + \text{hadrons})$$

and, through terms quadratic in J^{α}_{had}, purely non-leptonic interactions, such as:

$$\Lambda^o \rightarrow p + \pi^-$$

$$K^+ \rightarrow \pi^+ + \pi^+ + \pi^-$$

in which only hadrons appear. The coupling is weak in that the natural dimensionless coupling, with the proton mass as standard, $Gm_p^2 = 1.01 \times 10^{-5}$, where G is the Fermi constant. The leptonic current consists of the terms:

$$J^{\alpha}_{lept}(x) = \bar{e}\gamma^{\alpha}(1-\gamma_5)\nu_e + \bar{\mu}\gamma^{\alpha}(1-\gamma_5)\nu_{\mu} + \bar{\tau}\gamma^{\alpha}(1-\gamma_5)\nu_{\tau}$$

Both polar and axial vector terms appear ($\gamma_5 = i\gamma^0 \gamma^1 \gamma^2 \gamma^3$ is a pseudoscalar matrix), so that in the quadratic form $\mathcal{L}_W = -\frac{G}{\sqrt{2}} J^{\dagger}_{\alpha}(x) J^{\alpha}(x)$ there will be vector–axial-vector interference terms, indicating parity non-conservation. The discovery of this phenomenon, following the suggestion that reflection invariance in the weak interactions could not be taken for granted but had to be tested, played an important role in the determination of the phenomenological Lagrangian. The experiments suggested all involved looking for a pseudoscalar observable in a weak interaction experiment, and the first of many experiments measuring the beta decay of polarized nuclei (^{60}Co) showed an angular distribution of the form:

$$W(\theta) = A + Bp_e \cdot \langle J \rangle$$

where p_e is the electron momentum and $\langle j \rangle$ the polarization of the nucleus. The distribution $W(\theta)$ is not invariant under mirror inversion (P) which changes $J \rightarrow J$ and $p_e \rightarrow -p_e$, so the experimental form directly showed parity non-conservation. Experiments showed that both the hadronic and the leptonic currents had vector and axial-vector parts, and that although invariance under particle–antiparticle (charge) conjugation C is also violated, the form maintains invariance under the joint symmetry CP when restricted to the light hadrons (those consisting of u, d, c, and s). There is evidence that CP itself is violated at a much weaker level, of the order of 10^{-5} of the weak interactions. This is consistent with secondorder weak effects involving the heavy (b, t) quarks, though it is possible that an otherwise undetected superweak interaction also plays a role. The part of J^{α}_{had} relevant to beta decay is $\sim \bar{u}\gamma_{\alpha}(1-\gamma_5)d$. The detailed form of the hadronic current will be discussed after the description of the strong interactions.

Even at the leptonic level the theory described by $\mathcal{L}_W = -\frac{G}{\sqrt{2}} J^{\dagger}_{\alpha}(x) J^{\alpha}(x)$ is not renormalizable. This manifests itself in the result that the cross section for neutrino absorption grows with energy:

$$\sigma_{\nu} = (\text{const}) G^2 m_p E_{\nu}$$

While this behavior is in accord with observations up to the highest energies studied so far, it signals a breakdown of the theory at higher energies, so that $\mathcal{L}_W = -\frac{G}{\sqrt{2}} J^{\dagger}_{\alpha}(x) J^{\alpha}(x)$ cannot be fundamental. A number of people suggested over the years that the effective

Lagrangian is but a phenomenological description of a theory in which the weak current J α(x) is coupled to a charged intermediate vector boson $W_{\alpha}^{-}(x)$, in analogy with quantum electrodynamics. The form $\mathcal{L}_W = -\frac{G}{\sqrt{2}} J_{\alpha}^{\dagger}(x) J^{\alpha}(x)$ emerges from the exchange of a vector meson between the currents when the W mass is much larger than the momentum transfer in the process. The intermediate vector boson theory leads to a better behaved σ_v at high energies. However, massive vector theories are still not renormalizable, and the cross section for $e^+ e^- \rightarrow W^+W^-$ (with longitudinally polarized Ws) grows with energy. Until 1967 there was no theory of the weak interactions in which higher-order corrections, though extraordinarily small because of the weak coupling, could be calculated.

Classification

Classification of particles based on their spin and the interactions that they are subject to. Charged leptons, such as the electron and muon, are subject to the electromagnetic and weak interactions, while leptons without charge, i.e. neutrinos are subject to only the weak interaction. Hadrons, such as the proton and pions are subject to the electromagnetic, weak and strong interactions.

Bosons

In particle physics, a boson is a type of particle that obeys the rules of Bose-Einstein statistics. These bosons also have a quantum spin with contains an integer value, such as 0, 1, -1, -2, 2, etc. (By comparison, there are other types of particles, called fermions, that have a half-integer spin, such as 1/2, -1/2, -3/2, and so on).

Bosons are sometimes called force particles, because it is the bosons that control the interaction of physical forces, such as electromagnetism and possibly even gravity itself.

The name boson comes from the surname of Indian physicist Satyendra Nath Bose, a brilliant physicist from the early twentieth century who worked with Albert Einstein to develop a method of analysis called Bose-Einstein statistics. In an effort to fully understand Planck's law (the thermodynamics equilibrium equation that came out of Max Planck's work on the blackbody radiation problem), Bose first proposed the method in a 1924 paper trying to analyze the behavior of photons. He sent the paper to Einstein, who was able to get it published.

One of the most dramatic effects of Bose-Einstein statistics is the prediction that bosons can overlap and coexist with other bosons. Fermions, on the other hand, cannot do this, because they follow the Pauli Exclusion Principle (chemists focus primarily on the way the Pauli Exclusion Principle impacts the behavior of electrons in orbit around an atomic nucleus.) Because of this, it is possible for photons to become a laser and some matter is able to form the exotic state of a Bose-Einstein condensate.

Fundamental Bosons

There are a number of fundamental bosons, which are not made up of smaller particles. This includes the basic gauge bosons, the particles that mediate the fundamental forces of physics (except for gravity, which we'll get to in a moment). These four gauge bosons have spin 1 and have all been experimentally observed:

- Photon: Known as the particle of light, photons carry all electromagnetic energy and act as the gauge boson that mediates the force of electromagnetic interactions.
- Gluon: Gluons mediate the interactions of the strong nuclear force, which binds together quarks to form protons and neutrons and also holds the protons and neutrons together within an atom's nucleus.
- W Boson: One of the two gauge bosons involved in mediating the weak nuclear force.
- Z Boson: One of the two gauge bosons involved in mediating the weak nuclear force.

In addition to the above, there are other fundamental bosons predicted, but without clear experimental confirmation (yet):

- Higgs Boson: The Higgs Boson is the particle that gives rise to all mass. Scientists at the Large Hadron Collider announced that they had good reason to believe they'd found evidence of the Higgs Boson. Further research is ongoing in an attempt to get better information about the particle's exact properties. The particle is predicted to have a quantum spin value of 0, which is why it is classified as a boson.
- Graviton: The graviton is a theoretical particle which has not yet been experimentally detected. Since the other fundamental forces - electromagnetism, strong nuclear force, and weak nuclear force - are all explained in terms of a gauge boson that mediates the force, it was only natural to attempt to use the same mechanism to explain gravity. The resulting theoretical particle is the graviton, which is predicted to have a quantum spin value of 2.
- Bosonic Superpartners: Under the theory of super symmetry, every fermion would have a so-far-undetected bosonic counterpart. Since there are 12 fundamental fermions, this would suggest that - if super symmetry is true - there are another 12 fundamental bosons that have not yet been detected, presumably because they are highly unstable and have decayed into other forms.

Composite Bosons

Some bosons are formed when two or more particles join together to create an integer-spin particle, such as:

- Mesons: Mesons are formed when two quarks bond together. Since quarks are fermions and have half-integer spins, if two of them are bonded together, then the spin of the resulting particle (which is the sum of the individual spins) would be an integer, making it a boson.
- Helium-4 atom: A helium-4 atom contains 2 protons, 2 neutrons, and 2 electrons ... and if you add up all of those spins, you'll end up with an integer every time. Helium-4 is particularly noteworthy because it becomes a superfluid when cooled to ultra-low temperatures, making it a brilliant example of Bose-Einstein statistics in action.

Fermion

In particle physics, fermions are a group of elementary (or fundamental) particles that are the building blocks of matter. In the Standard Model, elementary particles are classified as fermions and bosons. Fermions are usually related with matter, whereas bosons are related with fundamental forces (or radiation).

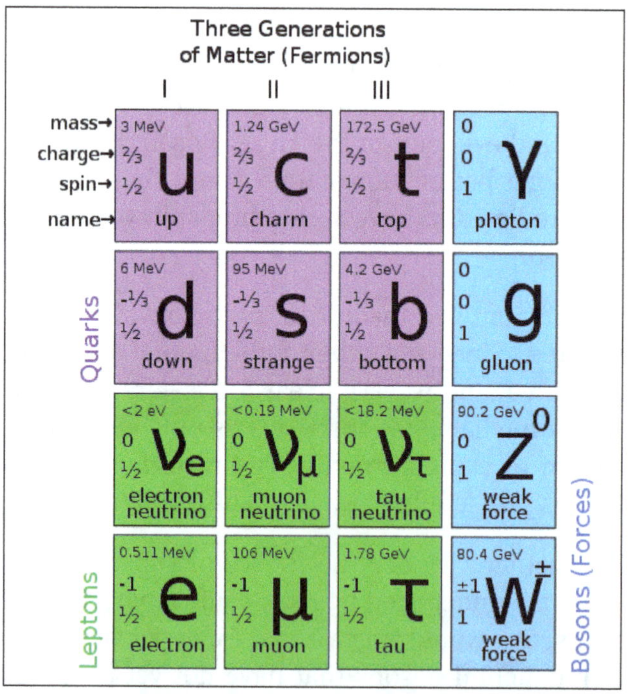

Standard Model of Elementary Particles.

Fermions are subdivided into quarks and leptons. Quarks are fermions that couple with a class of bosons known as gluons to form composite particles such as protons and neutrons. Leptons are those fermions that do not undergo coupling with gluons. Electrons are a well-known example of leptons.

Fermions come in pairs, and in three "generations." Everyday matter is composed of the first generation of fermions: two leptons, the electron and electron-neutrino; and two quarks, called Up and Down. Fermions obey what is known as "Fermi-Dirac statistics" and are named after Enrico Fermi.

Basic Properties

In theoretical terms, one major difference between fermions and bosons is related to a property known as "spin." Fermions have odd half-integer spin (1/2, 3/2, 5/2, and so forth), whereas bosons have integer spin (0, 1, 2, and so forth). Fermions obey Fermi-Dirac statistics, which means that when one swaps two fermions, the wavefunction of the system changes sign.

Given that each fermion has half-integer spin, when an observer circles a fermion (or when the fermion rotates 360° about its axis), the wavefunction of the fermion changes sign. A related phenomenon is called an antisymmetric wavefunction behavior of a fermion.

As particles with half-integer spin, fermions obey the Pauli exclusion principle: no two fermions can exist in the same quantum state at the same time. Thus, for more than one fermion to occupy the same place in space, certain properties (such as spin) of each fermion in the group must be different from the rest. The operation of the Pauli exclusion principle is used to explain the "rigidness" or "stiffness" of ordinary matter (contributing to the Young modulus of matter), and the stability of the electron shells of atoms (thus the stability of atomic matter). This principle is also responsible for the complexity of atoms (making it impossible for all atomic electrons to occupy the same energy level), thus making complex chemistry possible. In addition, this principle is said to be responsible for the pressure within degenerate matter, which largely governs the equilibrium state of white dwarfs and neutron stars.

In large systems, the difference between bosonic and fermionic statistics is apparent only at high densities, when their wave functions overlap. At low densities, both types of statistics are well approximated by Maxwell-Boltzmann statistics, which is described by classical mechanics.

Elementary Fermions

Elementary particles are classified as fermions and bosons, and elementary fermions are subdivided into quarks and leptons. When quarks are coupled together, they form composite fermions such as protons and neutrons. Leptons include the electron and similar, heavier particles (muon and tau) and neutrinos.

The known fermions of left-handed helicity interact through the weak interaction, but the known right-handed fermions do not. Or, put another way, only left-handed fermions and right-handed anti-fermions couple to the W boson.

Flavors of Fermions

There are 24 fundamental (or elementary) types of fermions, referred to as fermionic "flavors." They consist of 12 quarks and 12 leptons.

- 12 quarks - Six particles and six corresponding antiparticles:
  - Up (u)
  - Down (d)
  - Charm (c)
  - Strange (s)
  - Top (t)
  - Bottom (b)
  - Up antiquark (u)
  - Down antiquark (d)
  - Charm antiquark (c)
  - Strange antiquark (s)
  - Top antiquark (t)
  - Bottom antiquark (b)
- 12 leptons - Six particles and six corresponding antiparticles:
  - Electron (e^-)
  - Muon (μ^-)
  - Tau (τ^-)
  - Electron neutrino (ν_e)
  - Muon neutrino ($\nu\mu$)
  - Tau neutrino ($\nu\tau$)
  - Positron (e^+)

- Antimuon (μ^+)
- Antitau (τ^+)
- Electron antineutrino ($\overline{\nu}_e$)
- Muon antineutrino ($\overline{\nu}_\mu$)
- Tau antineutrino ($\overline{\nu}_\tau$)

Composite Fermions

In addition to elementary fermions and bosons, nonrelativistic composite particles made up of more fundamental particles bound together through a potential energy are composite fermions or bosons, depending only on the number of elementary fermions they contain:

- A composite particle containing an even number of elementary fermions is a boson. Examples:
  - A meson contains two fermion quarks and is a boson.
  - The nucleus of a carbon-12 atom contains six protons and six neutrons (all fermions) and is also a boson.
- A composite particle containing an odd number of elementary fermions is a fermion. Examples:
  - A baryon contains three quarks and is therefore a fermion.
  - The nucleus of a carbon-13 atom contains six protons and seven neutrons and is therefore a fermion.

The number of bosons within a composite particle made up of simple particles bound with a potential has no effect on whether the composite particle is a boson or a fermion. In a quantum field theory, the situation is more interesting. There can be field configurations of bosons that are topologically twisted. These are coherent states that behave like particles, and they can be fermionic even if all the elementary particles are bosons. This situation was discovered by Tony Skyrme in the early 1960s, so fermions made of bosons are named Skyrmions.

Fermionic or bosonic behavior of a composite particle (or system) is seen only at large distances (compared to the size of the system). At proximity, where spatial structure begins to be important, a composite particle (or system) behaves according to its constituent makeup. For example, two atoms of helium cannot share the same space if it is comparable by size to the size of the inner structure of the helium atom itself (~10−10 m)—despite bosonic properties of the helium atoms. Thus, liquid helium has finite density comparable to the density of ordinary liquid matter. The following table is based in part on data gathered by the Particle Data Group.

Table of fermions and their properties.

Generation 1						
Fermion (left-handed)	Symbol	Electric charge	Weak isospin	Weak hypercharge	Color charge	Mass
Electron	e^-	-1	-1/2	-1	1	511 keV
Positron	e^+	+1	0	+2	1	511 keV
Electron-neutrino	ν_e	0	+1/2	-1	1	< 2 eV
Up quark	u	+2/3	+1/2	+1/3	3	~ 3 MeV
Up antiquark	$\bar{u}$	-2/3	0	-4/3	$\bar{3}$	~ 3 MeV
Down quark	d	-1/3	-1/2	+1/3	3	~ 6 MeV
Down antiquark	$\bar{d}$	+1/3	0	+2/3	$\bar{3}$	~ 6 MeV

Generation 2						
Fermion (left-handed)	Symbol	Electric charge	Weak isospin	Weak hypercharge	Color charge	Mass
Muon	μ^-	-1	-1/2	-1	1	106 MeV
Antimuon	μ^+	+1	0	+2	1	106 MeV
Muon-neutrino	ν_μ	0	+1/2	-1	1	< 2 eV
Charm quark	c	+2/3	+1/2	+1/3	3	~ 1.337 GeV
Charm antiquark	$\bar{c}$	-2/3	0	-4/3	$\bar{3}$	~ 1.3 GeV
Strange quark	s	-1/3	-1/2	+1/3	3	~ 100 MeV
Strange antiquark	$\bar{s}$	+1/3	0	+2/3	$\bar{3}$	~ 100 MeV

Generation 3						
Fermion (left-handed)	Symbol	Electric charge	Weak isospin	Weak hypercharge	Color charge	Mass
Tau lepton	τ^-	-1	-1/2	-1	1	1.78 GeV
Anti-tau lepton	τ^+	+1	0	+2	1	1.78 GeV
Tau-neutrino	ν_τ	0	+1/2	-1	1	< 2 eV
Top quark	t	+2/3	+1/2	+1/3	3	171 GeV
Top antiquark	$\bar{t}$	-2/3	0	-4/3	$\bar{3}$	171 GeV
Bottom quark	b	-1/3	-1/2	+1/3	3	~ 4.2 GeV
Bottom antiquark	$\bar{b}$	+1/3	0	+2/3	$\bar{3}$	~ 4.2 GeV

Hypothetical Particles

Hypothetical particles, in Physics, are particles that have not yet been proved to exist by observation. However they have been or are being hypothesized about because of their occurrence in mathematical analysis in Physics. This is similar to the hypothesizing of the existence of Neptune in Newton's times. Supersymmetric theories predict the existence of more particles, none of which have been confirmed experimentally:

- The photino (spin-½) is the superpartner of the photon.
- The gluino (spin-½) is the superpartner of the gluon.
- The gravitino (spin-$^3/_2$) is the superpartner of the graviton boson in supergravity theories.
- The neutralino (spin-½) is a superposition of the superpartners of several neutral Standard Model particles. It is a leading candidate for dark matter. The partners of charged bosons are called charginos.
- Sterile neutrinos are introduced by many extensions to the Standard Model, and may be needed to explain the LSND results.
- Sleptons and squarks (spin-0) are the supersymmetric partners of the Standard Model fermions. The stop squark (superpartner of the top quark) is thought to have a low mass and is often the subject of experimental searches.

Other theories predict the existence of additional bosons:

- The Higgs (spin-0) has been proposed to explain the origin of mass by the spontaneous symmetry breaking of the SU(2) gauge symmetry.
- The graviton (spin-2) has been proposed to mediate gravity in theories of quantum gravity.
- The graviscalar (spin-0) and graviphoton (spin-1).
- The axion (spin-0) is a pseudoscalar particle introduced in Peccei-Quinn theory to solve the strong-CP problem.
- The axino and saxion form together with the axion a supermultiplet in supersymmetric extensions of Peccei-Quinn theory.
- The X and Y bosons are predicted by GUT theories to be heavier equivalents of the W and Z.
- The magnetic photon.

- The Majoron is predicted to understand neutrino masses by the seesaw mechanism.

Mirror particles are predicted by theories that restore Parity symmetry. Magnetic monopole is a generic name for particles with non-zero magnetic charge. They are predicted by some GUT theories. Tachyon is a generic name for hypothetical particles that travel faster than the speed of light and have an imaginary rest mass. The preon was a suggested substructure for both quarks and leptons, but modern collider experiments have all but disproven their existence.

References

- Elementary-Particles-in-Physics: physics.upenn.edu, Retrieved 03, July 2020
- Boson-2699112: thoughtco.com, Retrieved 12, March 2020
- Elementary-particles: chemeurope.com, Retrieved 28, January 2020

4

Quarks and Hadrons: An Integrated Study

Quarks are found within protons and neutrons, and are considered to be the building blocks of matter. The particles which experience the strong nuclear force are called hadrons. This chapter closely examines the key concepts of quarks and hadrons as well as color charge to provide an extensive understanding of the subject.

Quark

A quark is one of the fundamental particles in physics. They join to form hadrons, such as protons and neutrons, which are components of the nuclei of atoms. The study of quarks and the interactions between them through the strong force is called particle physics.

The antiparticle of a quark is the antiquark. Quarks and antiquarks are the only two fundamental particles that interact through all four fundamental forces of physics: gravitation, electromagnetism, and the strong and weak interactions.

Quarks and Confinement

A quark exhibits confinement, which means that the quarks are not observed independently but always in combination with other quarks. This makes determining the properties (mass, spin, and parity) impossible to measure directly; these traits must be inferred from the particles composed of them.

These measurements indicate a non-integer spin (either +1/2 or -1/2), so quarks are fermions and follow the Pauli Exclusion Principle.

In the strong interaction between quarks, they exchange gluons, which are massless vector gauge bosons that carry a pair of color and anticolor charges. When exchanging gluons, the color of the quarks change. This color force is weakest when the quarks are close together and becomes stronger as they move apart.

Quarks are so strongly bound by the color force that if there is enough energy to separate them, a quark-antiquark pair is produced and binds with any free quark to produce a hadron. As a result, free quarks are never seen alone.

Classification

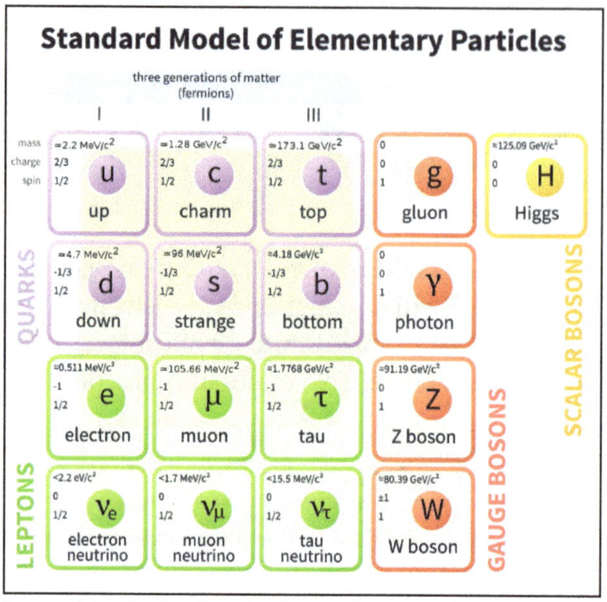

Six of the particles in the Standard Model are quarks (shown in purple).
Each of the first three columns forms a generation of matter.

The Standard Model is the theoretical framework describing all the currently known elementary particles. This model contains six flavors of quarks (q), named up (u), down (d), charm (c), strange (s), top (t), and bottom (b). Antiparticles of quarks are called antiquarks, and are denoted by a bar over the symbol for the corresponding quark, such as u for an up antiquark. As with antimatter in general, antiquarks have the same mass, mean lifetime, and spin as their respective quarks, but the electric charge and other charges have the opposite sign.

Quarks are spin-½ particles, implying that they are fermions according to the spin–statistics theorem. They are subject to the Pauli exclusion principle, which states that no two identical fermions can simultaneously occupy the same quantum state. This is in contrast to bosons (particles with integer spin), any number of which can be in the same state. Unlike leptons, quarks possess color charge, which causes them to engage in the strong interaction. The resulting attraction between different quarks causes the formation of composite particles known as hadrons.

The quarks that determine the quantum numbers of hadrons are called valence quarks; apart from these, any hadron may contain an indefinite number of virtual (or sea) quarks, antiquarks, and gluons, which do not influence its quantum numbers. There are two families of hadrons: baryons, with three valence quarks, and mesons, with a valence quark and an antiquark. The most common baryons are the proton and the neutron, the building blocks of the atomic nucleus. A great number of hadrons are known, most of them differentiated by their quark content and the properties these constituent quarks confer. The existence of "exotic" hadrons with more valence quarks, such as tetraquarks ($qq\bar{q}\bar{q}$) and pentaquarks ($qqqq\bar{q}$), were conjectured from the beginnings of the quark model but not discovered until the early 21st century.

Elementary fermions are grouped into three generations, each comprising two leptons and two quarks. The first generation includes up and down quarks, the second strange and charm quarks, and the third bottom and top quarks. All searches for a fourth generation of quarks and other elementary fermions have failed, and there is strong indirect evidence that no more than three generations exist. Particles in higher generations generally have greater mass and less stability, causing them to decay into lower-generation particles by means of weak interactions. Only first-generation (up and down) quarks occur commonly in nature. Heavier quarks can only be created in high-energy collisions (such as in those involving cosmic rays), and decay quickly; however, they are thought to have been present during the first fractions of a second after the Big Bang, when the universe was in an extremely hot and dense phase (the quark epoch). Studies of heavier quarks are conducted in artificially created conditions, such as in particle accelerators.

Having electric charge, mass, color charge, and flavor, quarks are the only known elementary particles that engage in all four fundamental interactions of contemporary physics: electromagnetism, gravitation, strong interaction, and weak interaction. Gravitation is too weak to be relevant to individual particle interactions except at extremes of energy (Planck energy) and distance scales (Planck distance). However, since no successful quantum theory of gravity exists, gravitation is not described by the Standard Model.

Flavors of Quarks

There are six flavors of quarks: up, down, strange, charm, bottom, and top. The flavor of the quark determines its properties.

Quarks with a charge of +(2/3)e are called up-type quarks and those with a charge of -(1/3)e are called down-type.

There are three generations of quarks, based on pairs of weak positive/negative, weak isospin. The first generation quarks are up and down quarks, the second generation quarks are strange, and charm quarks, the third generation quarks are top and bottom quarks.

All quarks have a baryon number (B = 1/3) and a lepton number (L = 0). The flavor determines certain other unique properties, described in individual descriptions.

The up and down quarks make up protons and neutrons, seen in the nucleus of ordinary matter. They are the lightest and most stable. The heavier quarks are produced in high-energy collisions and rapidly decay into up and down quarks. A proton is composed of two up quarks and a down quark. A neutron is composed of one up quark and two down quarks.

First-generation Quarks

Up quark (symbol *u*)

- Weak Isospin: +1/2.
- Isospin (I_z): +1/2.
- Charge (proportion of *e*): +2/3.
- Mass (in MeV/c^2): 1.5 to 4.0.

Down quark (symbol *d*)

- Weak Isospin: -1/2.
- Isospin (I_z): -1/2.
- Charge (proportion of *e*): -1/3.
- Mass (in MeV/c^2): 4 to 8.

Second-generation Quarks

Charm quark (symbol *c*)

- Weak Isospin: +1/2.
- Charm (*C*): 1.
- Charge (proportion of *e*): +2/3.
- Mass (in MeV/c^2): 1150 to 1350.

Strange quark (symbol *s*)

- Weak Isospin: -1/2.
- Strangeness (*S*): -1.
- Charge (proportion of *e*): -1/3.
- Mass (in MeV/c^2): 80 to 130.

Third-generation Quarks

Top quark (symbol *t*)

- Weak Isospin: +1/2.
- Topness (*T*): 1.
- Charge (proportion of *e*): +2/3.
- Mass (in MeV/c^2): 170200 to 174800.

Bottom quark (symbol *b*)

- Weak Isospin: -1/2.
- Bottomness (*B′*): 1.
- Charge (proportion of *e*): -1/3.
- Mass (in MeV/c^2): 4100 to 4400.

Properties

Electric Charge

Quarks have fractional electric charge values – either (−⅓) or (+⅔) times the elementary charge (e), depending on flavor. Up, charm, and top quarks (collectively referred to as up-type quarks) have a charge of +⅔ e, while down, strange, and bottom quarks (down-type quarks) have −⅓ e. Antiquarks have the opposite charge to their corresponding quarks; up-type antiquarks have charges of −⅔ e and down-type antiquarks have charges of +⅓ e. Since the electric charge of a hadron is the sum of the charges of the constituent quarks, all hadrons have integer charges: the combination of three quarks (baryons), three antiquarks (antibaryons), or a quark and an antiquark (mesons) always results in integer charges. For example, the hadron constituents of atomic nuclei, neutrons and protons, have charges of 0 e and +1 e respectively; the neutron is composed of two down quarks and one up quark, and the proton of two up quarks and one down quark.

Spin

Spin is an intrinsic property of elementary particles, and its direction is an important degree of freedom. It is sometimes visualized as the rotation of an object around its own axis (hence the name "spin"), though this notion is somewhat misguided at subatomic scales because elementary particles are believed to be point-like.

Spin can be represented by a vector whose length is measured in units of the reduced Planck constant ħ (pronounced "h bar"). For quarks, a measurement of the spin vector component along any axis can only yield the values +ħ/2 or −ħ/2; for this reason quarks are classified as spin-½ particles. The component of spin along a given axis – by convention the z axis – is often denoted by an up arrow ↑ for the value +½ and down arrow ↓ for the value −½, placed after the symbol for flavor. For example, an up quark with a spin of +½ along the z axis is denoted by u↑.

Weak Interaction

A tree diagram consisting mostly of straight arrows. A down quark forks into an up quark and a wavy-arrow W^- boson, the latter forking into an electron and reversed-arrow electron antineutrino.

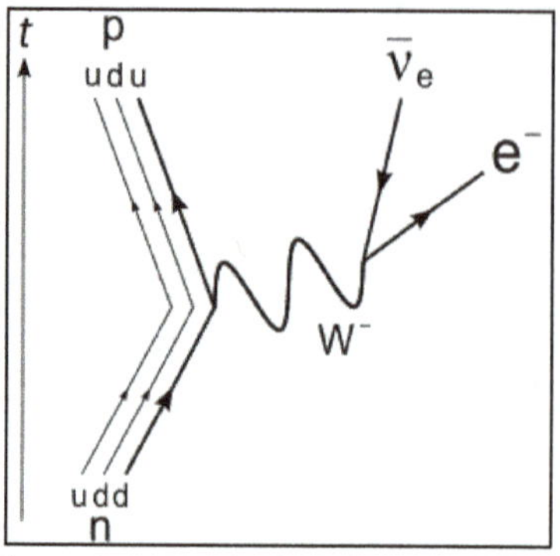

Feynman diagram of beta decay with time flowing upwards. The CKM matrix encodes the probability of this and other quark decays.

A quark of one flavor can transform into a quark of another flavor only through the weak interaction, one of the four fundamental interactions in particle physics. By absorbing or emitting a W boson, any up-type quark (up, charm, and top quarks) can change into any down-type quark (down, strange, and bottom quarks) and vice versa. This flavor transformation mechanism causes the radioactive process of beta decay, in which a neutron (n) "splits" into a proton (p), an electron (e–) and an electron antineutrino ($\bar{\nu}_e$). This occurs when one of the down quarks in the neutron (udd) decays into an up quark by emitting a virtual W^- boson, transforming the neutron into a proton (uud). The W– boson then decays into an electron and an electron antineutrino.

$$n \rightarrow p + e^- + \bar{\nu}_e \text{ (Beta decay, hadron notation)}$$

$$udd \rightarrow uud + e^- + \bar{\nu}_e \text{ (Beta decay, quark notation)}$$

Both beta decay and the inverse process of inverse beta decay are routinely used in medical applications such as positron emission tomography (PET) and in experiments involving neutrino detection.

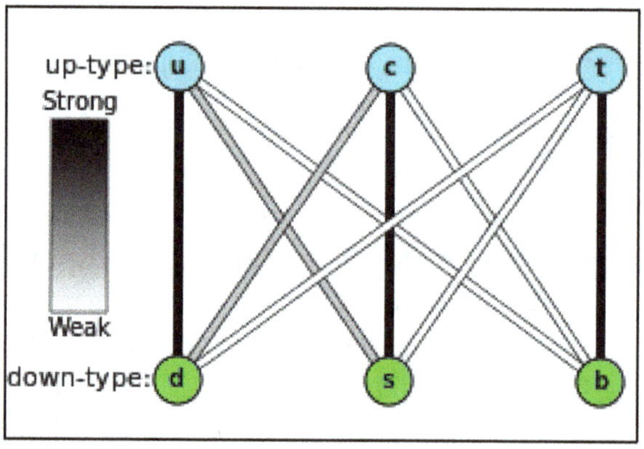

The strengths of the weak interactions between the six quarks. The "intensities" of the lines are determined by the elements of the CKM matrix.

While the process of flavor transformation is the same for all quarks, each quark has a preference to transform into the quark of its own generation. The relative tendencies of all flavor transformations are described by a mathematical table, called the Cabibbo–Kobayashi–Maskawa matrix (CKM matrix). Enforcing unitarity, the approximate magnitudes of the entries of the CKM matrix are:

$$\begin{bmatrix} |V_{ud}| & |V_{us}| & |V_{ub}| \\ |V_{cd}| & |V_{cs}| & |V_{cb}| \\ |V_{td}| & |V_{ts}| & |V_{tb}| \end{bmatrix} \approx \begin{bmatrix} 0.974 & 0.225 & 0.003 \\ 0.225 & 0.973 & 0.041 \\ 0.009 & 0.040 & 0.999 \end{bmatrix},$$

where V_{ij} represents the tendency of a quark of flavor i to change into a quark of flavor j (or vice versa).

There exists an equivalent weak interaction matrix for leptons, called the Pontecorvo–Maki–Nakagawa–Sakata matrix (PMNS matrix). Together, the CKM and PMNS matrices describe all flavor transformations, but the links between the two are not yet clear.

Strong Interaction and Color Charge

A green and a magenta ("antigreen") arrow canceling out each other out white, representing a meson; a red, a green, and a blue arrow canceling out to white, representing a baryon; a yellow ("antiblue"), a magenta, and a cyan ("antired") arrow canceling out to white, representing an antibaryon.

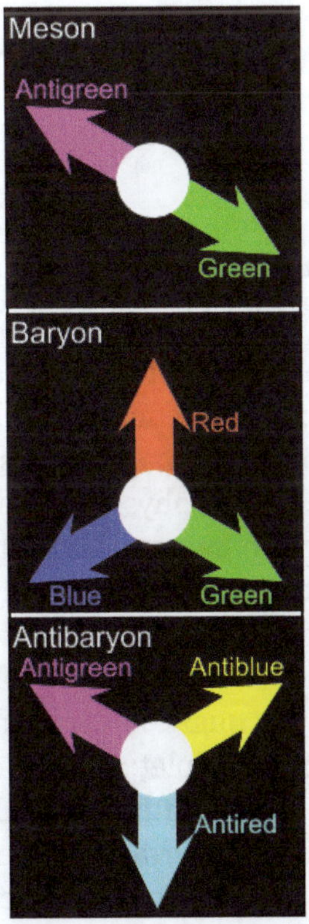

All types of hadrons have zero total color charge.

According to quantum chromodynamics (QCD), quarks possess a property called color charge. There are three types of color charge, arbitrarily labeled blue, green, and red. Each of them is complemented by an anticolor – antiblue, antigreen, and antired. Every quark carries a color, while every antiquark carries an anticolor.

The system of attraction and repulsion between quarks charged with different combinations of the three colors is called strong interaction, which is mediated by force carrying particles known as gluons; this is discussed at length below. The theory that describes strong interactions is called quantum chromodynamics (QCD). A quark, which will have a single color value, can form a bound system with an antiquark carrying the corresponding anticolor. The result of two attracting quarks will be color neutrality: a quark with color charge ξ plus an antiquark with color charge −ξ will result in a color charge of 0 (or "white" color) and the formation of a meson. This is analogous to the additive color model in basic optics. Similarly, the combination of three quarks, each with different color charges, or three antiquarks, each with anticolor charges, will result in the same "white" color charge and the formation of a baryon or antibaryon.

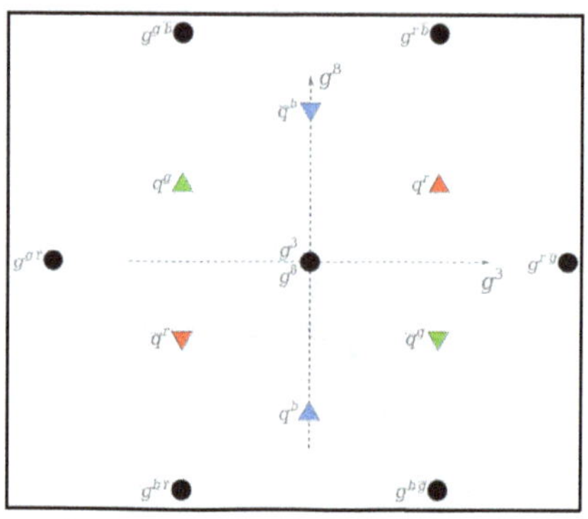

The pattern of strong charges for the three colors of quark, three antiquarks, and eight gluons (with two of zero charge overlapping).

In modern particle physics, gauge symmetries – a kind of symmetry group – relate interactions between particles. Color SU(3) (commonly abbreviated to $SU(3)_c$) is the gauge symmetry that relates the color charge in quarks and is the defining symmetry for quantum chromodynamics. Just as the laws of physics are independent of which directions in space are designated x, y, and z, and remain unchanged if the coordinate axes are rotated to a new orientation, the physics of quantum chromodynamics is independent of which directions in three-dimensional color space are identified as blue, red, and green. $SU(3)_c$ color transformations correspond to "rotations" in color space (which, mathematically speaking, is a complex space). Every quark flavor f, each with subtypes f_B, f_G, f_R corresponding to the quark colors, forms a triplet: a three-component quantum field which transforms under the fundamental representation of $SU(3)_c$. The requirement that $SU(3)_c$ should be local – that is, that its transformations be allowed to vary with space and time – determines the properties of the strong interaction. In particular, it implies the existence of eight gluon types to act as its force carriers.

Mass

Two terms are used in referring to a quark's mass: current quark mass refers to the mass of a quark by itself, while constituent quark mass refers to the current quark mass plus the mass of the gluon particle field surrounding the quark. These masses typically have very different values. Most of a hadron's mass comes from the gluons that bind the constituent quarks together, rather than from the quarks themselves. While gluons are inherently massless, they possess energy – more specifically, quantum chromodynamics binding energy (QCBE) – and it is this that contributes so greatly to the overall mass of the hadron. For example, a proton has a mass of approximately 938 MeV/c^2, of which the rest mass of its three valence quarks only contributes about 9 MeV/c^2; much of the remainder can be attributed to the field energy of the gluons.

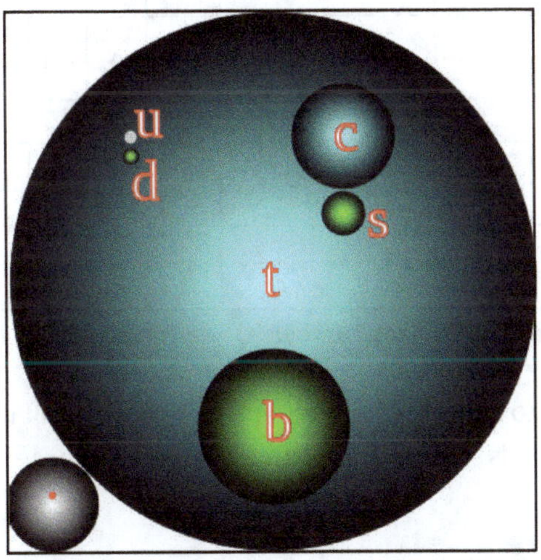

Current quark masses for all six flavors in comparison, as balls of proportional volumes. Proton (gray) and electron (red) are shown in bottom left corner for scale.

The Standard Model posits that elementary particles derive their masses from the Higgs mechanism, which is associated to the Higgs boson. It is hoped that further research into the reasons for the top quark's large mass of ~173 GeV/c^2, almost the mass of a gold atom, might reveal more about the origin of the mass of quarks and other elementary particles.

Table of Properties

The following table summarizes the key properties of the six quarks. Flavor quantum numbers (isospin (I_3), charm (C), strangeness, topness (T), and bottomness (B′)) are assigned to certain quark flavors, and denote qualities of quark-based systems and hadrons. The baryon number (B) is +1⁄3 for all quarks, as baryons are made of three quarks. For antiquarks, the electric charge (Q) and all flavor quantum numbers (B, I_3, C, S, T, and B′) are of opposite sign. Mass and total angular momentum (J; equal to spin for point particles) do not change sign for the antiquarks.

Quark Flavor Properties

Name	Symbol	Mass (MeV/c²)*	J	B	Q (e)	I_3	C	S	T	B'	Antiparticle	Antiparticle symbol
First generation												
Up	u	2.3±0.7 ± 0.5	$^1/_2$	$+^1/_3$	$+^2/_3$	$+^1/_2$	0	0	0	0	Antiup	$\bar{u}$
Down	d	4.8±0.5 ± 0.3	$^1/_2$	$+^1/_3$	$-^1/_3$	$-^1/_2$	0	0	0	0	Antidown	$\bar{d}$
Second generation												
Charm	C	1275±25	½	$+^1/_3$	$+^2/_3$	0	+1	0	0	0	Antich-arm	$\bar{c}$
Strange	S	95±5	½	$+^1/_3$	$-^1/_3$	0	0	−1	0	0	An-tistrange	$\bar{s}$
Third generation												
Top	t	173210±510 ± 710	½	$+^1/_3$	$+^2/_3$	0	0	0	+1	0	Antitop	$\bar{t}$
Bottom	b	4180±30	½	$+^1/_3$	$-^1/_3$	0	0	0	0	−1	Antibot-tom	$\bar{b}$

J = total angular momentum, B = baryon number, Q = electric charge, I_3 = isospin, C = charm, S = strangeness, T = topness, B′ = bottomness.

* Notation such as 173210±510 ± 710 denotes two types of measurement uncertainty. In the case of the top quark, the first uncertainty is statistical in nature, and the second is systematic.

Interacting Quarks

As described by quantum chromodynamics, the strong interaction between quarks is mediated by gluons, massless vector gauge bosons. Each gluon carries one color charge and one anticolor charge. In the standard framework of particle interactions (part of a more general formulation known as perturbation theory), gluons are constantly exchanged between quarks through a virtual emission and absorption process. When a gluon is transferred between quarks, a color change occurs in both; for example, if a red quark emits a red–antigreen gluon, it becomes green, and if a green quark absorbs a red–antigreen gluon, it becomes red. Therefore, while each quark's color constantly changes, their strong interaction is preserved.

Since gluons carry color charge, they themselves are able to emit and absorb other gluons. This causes asymptotic freedom: as quarks come closer to each other, the chromodynamic binding force between them weakens. Conversely, as the distance between quarks increases, the binding force strengthens. The color field becomes stressed, much as an elastic band is stressed when stretched, and more gluons of appropriate color are spontaneously created to strengthen the field. Above a certain energy threshold, pairs of quarks and antiquarks are created. These pairs bind with the quarks being separated, causing new hadrons to form. This phenomenon is known as color confinement: quarks never appear in isolation. This process of hadronization occurs before quarks, formed in a high energy collision, are able to interact in any other way. The only exception is the top quark, which may decay before it hadronizes.

Sea Quarks

Hadrons contain, along with the valence quarks (q_v) that contribute to their quantum numbers, virtual quark–antiquark ($q\bar{q}$) pairs known as sea quarks (q_s). Sea quarks form when a gluon of the hadron's color field splits; this process also works in reverse in that the annihilation of two sea quarks produces a gluon. The result is a constant flux of gluon splits and creations colloquially known as "the sea". Sea quarks are much less stable than their valence counterparts, and they typically annihilate each other within the interior of the hadron. Despite this, sea quarks can hadronize into baryonic or mesonic particles under certain circumstances.

Other Phases of Quark Matter

Quark–gluon plasma exists at very high temperatures; the hadronic phase exists at lower temperatures and baryonic densities, in particular nuclear matter for relatively low temperatures and intermediate densities; color superconductivity exists at sufficiently low temperatures and high densities.

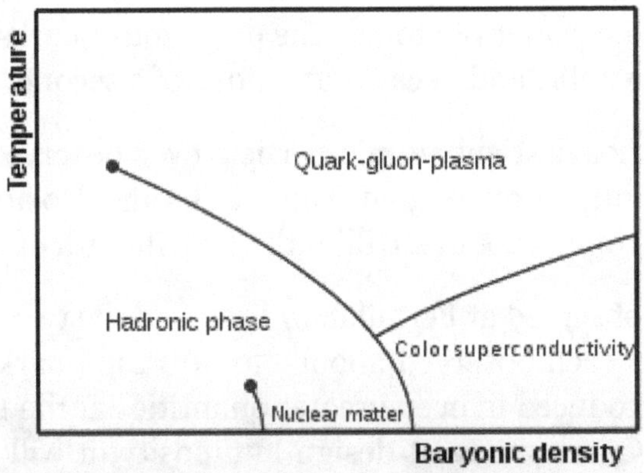

A qualitative rendering of the phase diagram of quark matter. The precise details of the diagram are the subject of ongoing research.

Under sufficiently extreme conditions, quarks may become deconfined and exist as free particles. In the course of asymptotic freedom, the strong interaction becomes weaker at higher temperatures. Eventually, color confinement would be lost and an extremely hot plasma of freely moving quarks and gluons would be formed. This theoretical phase of matter is called quark–gluon plasma. The exact conditions needed to give rise to this state are unknown and have been the subject of a great deal of speculation and experimentation. A recent estimate puts the needed temperature at $(1.90 \pm 0.02) \times 10^{12}$ kelvin.

While a state of entirely free quarks and gluons has never been achieved (despite numerous attempts by CERN in the 1980s and 1990s), recent experiments at the Relativistic Heavy Ion Collider have yielded evidence for liquid-like quark matter exhibiting "nearly perfect" fluid motion.

The quark–gluon plasma would be characterized by a great increase in the number of heavier quark pairs in relation to the number of up and down quark pairs. It is believed that in the period prior to 10–6 seconds after the Big Bang (the quark epoch), the universe was filled with quark–gluon plasma, as the temperature was too high for hadrons to be stable.

Given sufficiently high baryon densities and relatively low temperatures – possibly comparable to those found in neutron stars – quark matter is expected to degenerate into a Fermi liquid of weakly interacting quarks. This liquid would be characterized by a condensation of colored quark Cooper pairs, thereby breaking the local $SU(3)_c$ symmetry. Because quark Cooper pairs harbor color charge, such a phase of quark matter would be color superconductive; that is, color charge would be able to pass through it with no resistance.

Top Quark

As its name states, the top quark is a quark. Quarks are one of the fundamental building blocks of which the entire universe is constructed. Scientists know of six different quarks, although only two of them (the up and down quarks) are found inside the protons and neutrons at the center of atoms. The other four quarks (strange, charm, bottom and top) are all unstable and decay in fractions of a second.

The top quark is the heaviest subatomic particle ever observed, with a mass that is about as heavy as an entire atom of gold. Top quarks are also among the most fleeting of particles, with a lifetime of about a trillionth of a trillionth of a second.

Top quarks were first observed at Fermilab in 1995 and the two experiments at Fermilab's Tevatron collider each observed about 150,000 top quarks over their lifetimes. Top quarks are now produced in even greater quantities at the Large Hadron Collider (LHC). When the LHC is running at design luminosity, it will produce a pair of top quarks every second. As of June 2012, over two million top quarks have been created in each of the big experiments at the LHC, and the rate is rapidly increasing.

The original studies concentrated on collisions in which a top quark and an antimatter top quark were simultaneously produced. Experiments at Fermilab were also the first to observe top quarks produced one at a time. This was a tricky thing to accomplish because ordinary processes can produce top quark lookalikes. Making two lookalikes at the same time is rare, which makes it easier to separate real top quark events from fakes. To find singly made top quarks required a detailed understanding of the detector.

Though 17 years have passed since the top quark was discovered, it remains a topic of intense research interest. Perhaps the most pressing mystery in contemporary particle physics is the origin of mass. While the mechanism that gives particles their mass is not yet known, a very popular theory is that the Higgs boson is responsible. Given that the top quark is the heaviest particle yet discovered, it is a natural laboratory in which to investigate how the Higgs boson interacts with matter. Further, if the Higgs boson is real, the theory makes very precise predictions on how the mass of the Higgs boson and the top quark are related. In fact, one of the ways we will be certain that an observation of a Higgs boson is the real Higgs boson (as opposed to a competitor theory) will be to measure the mass of the top quark with sufficient precision to see if the prediction is correct.

After years of intense studies, the top quark has revealed a mystery or two of its own. At Fermilab's Tevatron, measurements by both of its collider experiments showed that the top quark prefers to be created in the same direction the proton beam is travelling. This measurement disagrees with predictions and could indicate new physical phenomena or it could just mean that the regular theory needs to be tweaked.

In addition, top quarks, produced singly or in pairs, have rich panoply of decay products with signatures that can mimic collisions of even rarer phenomena. Thus it is imperative to understand in great detail how top quarks are produced.

The top quark is an intriguing particle, massive and fleeting, and one that continues to provide a broad range of interesting research topics.

Properties

- At the final Tevatron energy of 1.96 TeV, top–antitop pairs were produced with a cross section of about 7 picobarns (pb). The Standard Model prediction (at next-to-leading order with m_t = 175 GeV/c^2) is 6.7–7.5 pb.
- The W bosons from top quark decays carry polarization from the parent particle, hence pose themselves as a unique probe to top polarization.
- In the Standard Model, the top quark is predicted to have a spin quantum number of $\frac{1}{2}$ and electric charge $+\frac{2}{3}$. A first measurement of the top quark charge has been published, resulting in approximately 90% confidence limit that the top quark charge is indeed $+\frac{2}{3}$.

Production

Because top quarks are very massive, large amounts of energy are needed to create one. The only way to achieve such high energies is through high energy collisions. These occur naturally in the Earth's upper atmosphere as cosmic rayscollide with particles in the air, or can be created in a particle accelerator. In 2011, after the Tevatron ceased operations, the Large Hadron Collider at CERN became the only accelerator that generates a beam of sufficient energy to produce top quarks, with a center-of-mass energy of 7 TeV. There are multiple processes that can lead to the production of top quarks, but they can be conceptually divided in two categories.

Top-quark Pairs

The most common is production of a top–antitop pair via strong interactions. In a collision, a highly energetic gluon is created, which subsequently decays into a top and antitop. This process was responsible for the majority of the top events at Tevatron and was the process observed when the top was first discovered in 1995. It is also possible to produce pairs of top–antitop through the decay of an intermediate photon or Z-boson. However, these processes are predicted to be much rarer and have a virtually identical experimental signature in a hadron collider like Tevatron.

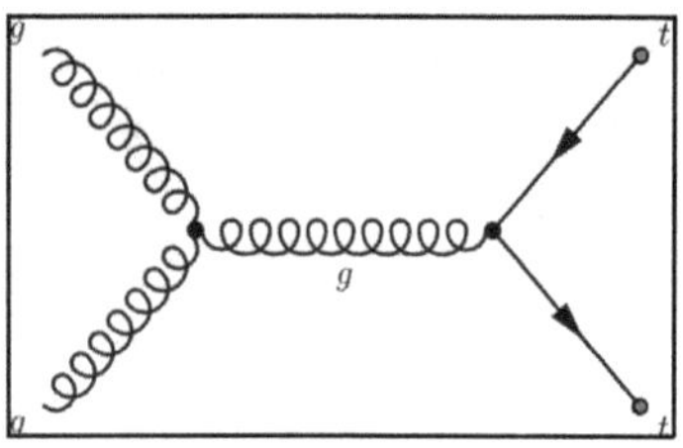

gluon-gluon fusion

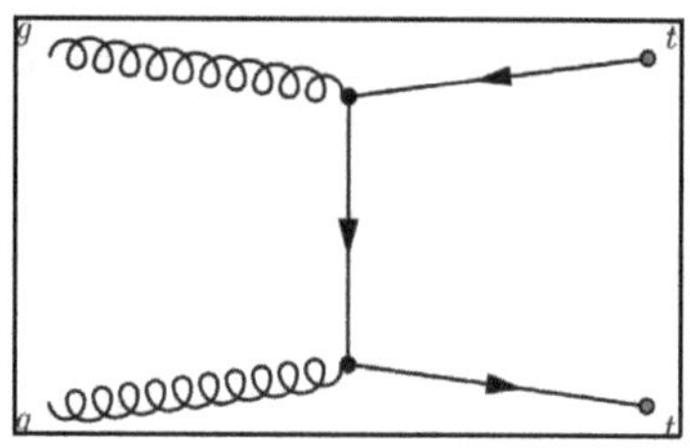

t-channel

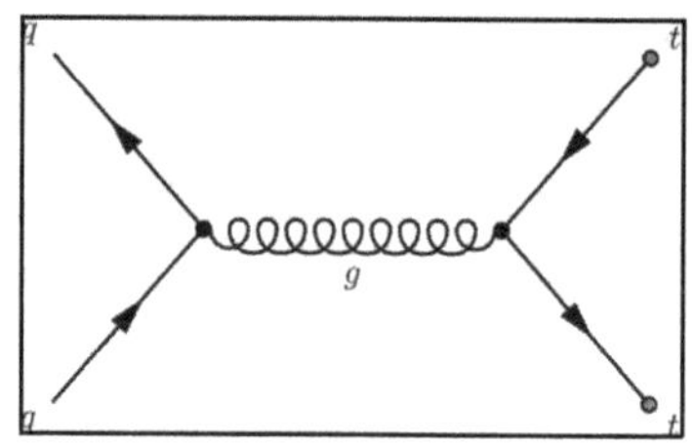

quark-antiquark annihilation

Single Top Quarks

A distinctly different process is the production of single top quarks via weak interaction. This can happen in several ways (called channels): either an intermediate W-boson decays into a top and antibottom quark ("s-channel") or a bottom quark (probably created in a pair through the decay of a gluon) transforms to a top quark by exchanging a W-boson with an up or down quark ("t-channel"). A single top quark can also be produced in association with a W boson, requiring an initial state bottom quark ("tW-channel"). The first evidence for these processes was published by the DØ collaboration in December 2006, and in March 2009 the CDF and DØ collaborations released twin papers with the definitive observation of these processes. The main significance of measuring these production processes is that their frequency is directly proportional to the $|V_{tb}|^2$ component of the CKM matrix.

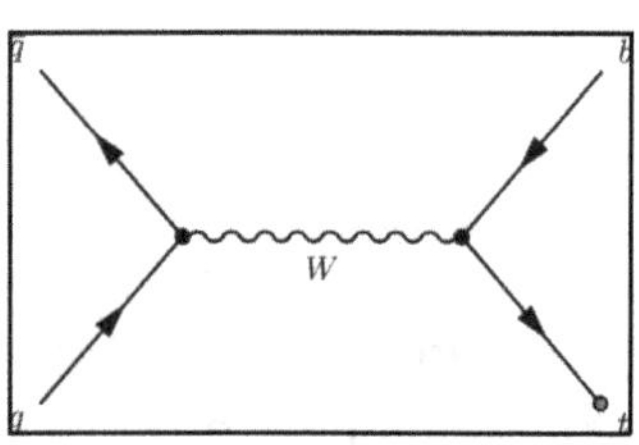

s-channel

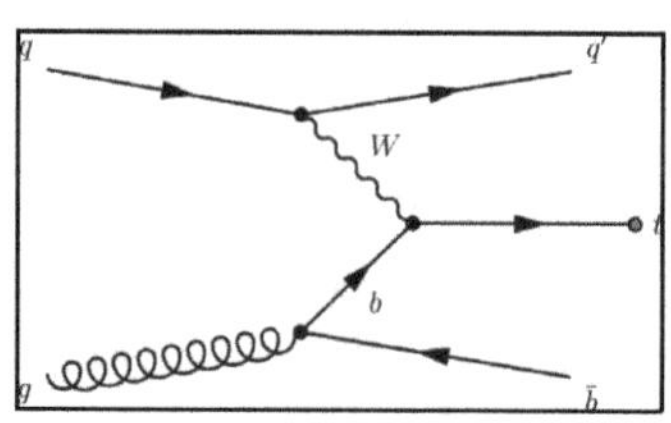

t-channel

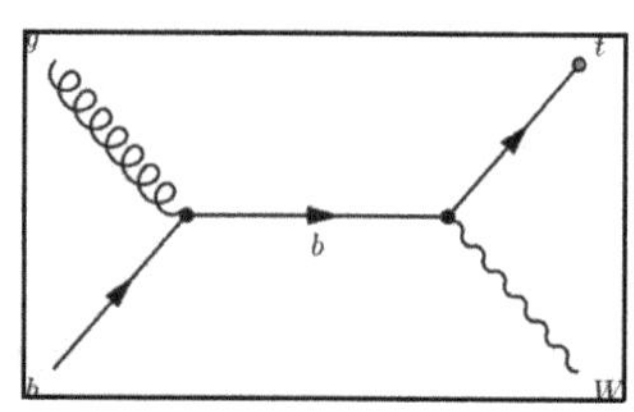

tW channel

Decay

Because of its enormous mass, the top quark is extremely short-lived with a predicted lifetime of only 5×10^{-25} s. As a result, top quarks do not have time before they decay to form hadrons as other quarks do, which provides physicists with the unique opportunity to study the behavior of a "bare" quark. The only known way the top quark can decay is through the weak interaction producing a W-boson and a down-type quark (down, strange, or bottom).

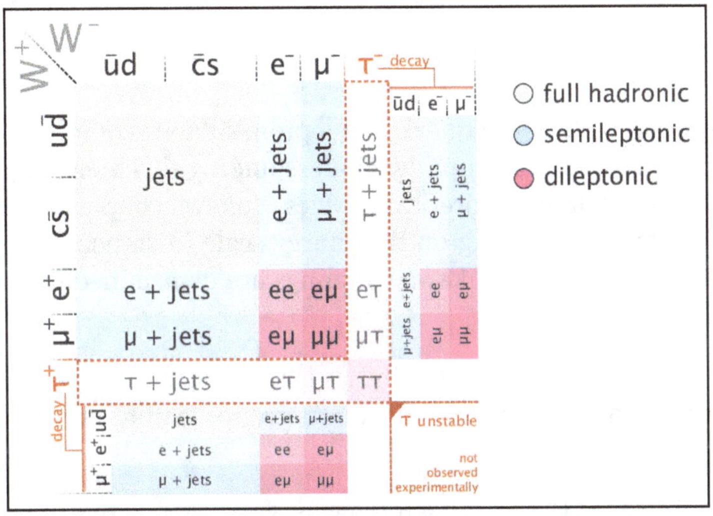

All possible final states of the decay of a top-quark pair.

In particular, it is possible to directly determine the branching ratio $\Gamma(W^+b)/\Gamma(W^+q\ (q = b,s,d))$. The best current determination of this ratio is 0.91±0.04. Since this ratio is equal to $|V_{tb}|^2$ according to the Standard Model, this gives another way of determining the CKM element $|V_{tb}|$, or in combination with the determination of $|V_{tb}|$ from single top production provides tests for the assumption that the CKM matrix is unitary.

The Standard Model also allows more exotic decays, but only at one loop level, meaning that they are extremely suppressed. In particular, it is possible for a top quark to decay into another up-type quark (an up or a charm) by emitting a photon or a Z-boson. Searches for these exotic decay modes have provided no evidence for their existence in accordance with expectations from the Standard Model. The branching ratios for these decays have been determined to be less than 5.9 in 1,000 for photonic decay and less than 2.1 in 1,000 for Z-boson decay at 95% confidence.

Mass and Coupling to the Higgs Boson

The Standard Model describes fermion masses through the Higgs mechanism. The Higgs boson has a Yukawa coupling to the left- and right-handed top quarks. After electroweak symmetry breaking (when the Higgs acquires a vacuum expectation value), the left- and right-handed components mix, becoming a mass term.

$$\mathcal{L} = y_t hqu^c \rightarrow \frac{y_t v}{\sqrt{2}}(1 + h^0 / v)uu^c$$

The top quark Yukawa coupling has a value of

$$y_t = \sqrt{2}m_t / v \simeq 1$$

where v = 246 GeV is the value of the Higgs vacuum expectation value.

Yukawa Couplings

In the Standard Model, all of the quark and lepton Yukawa couplings are small compared to the top quark Yukawa coupling. Understanding this hierarchy in the fermion masses is an open problem in theoretical physics. Yukawa couplings are not constants and their values change depending on the energy scale (distance scale) at which they are measured. The dynamics of Yukawa couplings are determined by the renormalization group equation.

One of the prevailing views in particle physics is that the size of the top quark Yukawa coupling is determined by the renormalization group, leading to the "quasi-infrared fixed point."

The Yukawa couplings of the up, down, charm, strange and bottom quarks, are hypothesized to have small values at the extremely high energy scale of grand unification, 10^{15} GeV. They increase in value at lower energy scales, at which the quark masses are generated by the Higgs. The slight growth is due to corrections from the QCD coupling. The corrections from the Yukawa couplings are negligible for the lower mass quarks.

If, however, a quark Yukawa coupling has a large value at very high energies, its Yukawa corrections will evolve and cancel against the QCD corrections. This is known as a (quasi-) infrared fixed point. No matter what the initial starting value of the coupling is, if it is sufficiently large it will reach this fixed point value. The corresponding quark mass is then predicted.

The top quark Yukawa coupling lies very near the infrared fixed point of the Standard Model. The renormalization group equation is:

$$\mu \frac{\partial}{\partial \mu} y_t \approx \frac{y_t}{16\pi^2}\left(\frac{9}{2}y_t^2 - 8g_3^2 - \frac{9}{4}g_2^2 - \frac{17}{20}g_1^2\right),$$

where g_3 is the color gauge coupling, g_2 is the weak isospin gauge coupling, and g_1 is the weak hypercharge gauge coupling. This equation describes how the Yukawa coupling changes with energy scale μ. Solutions to this equation for large initial values y_t cause the right-hand side of the equation to quickly approach zero, locking y_t to the QCD coupling g_3. The value of the fixed point is fairly precisely determined in the Standard Model, leading to a top quark mass of 230 GeV. However, if there is more than one Higgs doublet, the mass value will be reduced by Higgs mixing angle effects in an unpredicted way.

In the minimal supersymmetric extension of the Standard Model (MSSM), there are two Higgs doublets and the renormalization group equation for the top quark Yukawa coupling is slightly modified:

$$\mu\frac{\partial}{\partial\mu}y_{\mathrm{t}} \approx \frac{y_{\mathrm{t}}}{16\pi^2}\left(6y_{\mathrm{t}}^2 + y_{\mathrm{b}}^2 - \frac{16}{3}g_3^2 - 3g_2^2 - \frac{13}{15}g_1^2\right),$$

where y_{b} is the bottom quark Yukawa coupling. This leads to a fixed point where the top mass is smaller, 170–200 GeV. The uncertainty in this prediction arises because the bottom quark Yukawa coupling can be amplified in the MSSM. Some theorists believe this is supporting evidence for the MSSM.

The quasi-infrared fixed point has subsequently formed the basis of top quark condensation theories of electroweak symmetry breaking in which the Higgs boson is composite at *extremely* short distance scales, composed of a pair of top and antitop quarks.

Stop Squark

In particle physics, a stop squark, symbol $\tilde{t}$, is the superpartner of the top quark as predicted by supersymmetry (SUSY). It is a sfermion, which means it is a spin-0 boson (scalar boson). While the top quark is the heaviest known quark, the stop squark is actually often the lightest squark in many supersymmetry models.

The stop squark is a key ingredient of a wide range of SUSY models that address the hierarchy problem of the Standard Model (SM) in a natural way. A boson partner to the top quark would stabilize the Higgs boson mass against quadratically divergent quantum corrections, provided its mass is close to the electroweak symmetry breaking energy scale. If this was the case then the stop squark would be accessible at the Large Hadron Collider. In the generic R-parity conserving Minimal Supersymmetric Standard Model (MSSM) the scalar partners of right-handed and left-handed top quarks mix to form two stop mass eigenstates. Depending on the specific details of the SUSY model and the mass hierarchy of the sparticles, the stop might decay into a bottom quark and a chargino, with a subsequent decay of the chargino into the lightest neutralino (which is often the lightest supersymmetric particle).

Many searches for evidence of the stop squark have been performed by both the ATLAS and CMS experiments at the LHC but so far no signal has been discovered.

Bottom Quark

Bottom quarks (also called beauty quarks) are not stable and quickly decay, mostly to charmquarks and W^- bosons. In their short lifetime, they are part of some baryons and mesons. Due to their high mass, bottom quarks are relatively easy to research in experiments, a bottom quark is slightly heavier than a helium atom.

In 1977, an experimental group at Fermilab led by Leon Lederman discovered a new resonance at 9.4 GeV/c^2 which was interpreted as a bottom-antibottom quark pair and called the Upsilon meson. From this experiment, the mass of the bottom quark is implied to be about 5 GeV/c^2. The reaction being studied was,

$$p + N \rightarrow \mu^+ + \mu^- + X$$

where N was a copper or platinum nucleus. The spectrometer had a muon-pair mass resolution of about 2%, which allowed them to measure an excess of events at 9.4 GeV/c^2. This resonance has been subsequently studied at other accelerators with a detailed investigation of the bound states of the bottom-antibottom meson. Studies of the B-meson have also been productive.

Distinct Character

The bottom quark's "bare" mass is around 4.2 GeV/c^2 – a bit more than four times the mass of a proton, and many orders of magnitude larger than common "light" quarks.

Although it almost-exclusively transitions from or to a top quark, the bottom quark can decay into either an up quark or charm quark via the weak interaction. CKM matrixelements V_{ub} and V_{cb} specify the rates, where both these decays are suppressed, making lifetimes of most bottom particles (~10^{-12} s) somewhat higher than those of charmed particles (~10^{-13} s), but lower than those of strange particles (from ~10^{-10} to ~10^{-8} s).

The combination of high mass and low transition-rate gives experimental collisionby-products containing a bottom quark a distinctive signature that makes them relatively easy to identify using a technique called "B-tagging". For that reason, mesonscontaining the bottom quark are exceptionally long-lived for their mass, and are the easiest particles to use to investigate CP violation. Such experiments are being performed at th e BaBar, Belle and LHCb experiments.

Hadrons Containing Bottom Quarks

Some of the hadrons containing bottom quarks include:

- B mesons contain a bottom quark (or its antiparticle) and an up or down quark.
- B_c and B_s mesons contain a bottom quark along with a charm quark or strange quark respectively.

- There are many bottomonium states, for example the ϒ meson and $\chi_b(3P)$, the first particle discovered in LHC. These consist of a bottom quark and its antiparticle.
- Bottom baryons have been observed, and are named in analogy with strange baryons (e.g. Λ_b^0).

Up and Down Quark

The up and down quarks are the most common and least massive quarks, being the constituents of protons and neutrons and thus of most ordinary matter.

The fact that the free neutron decays

$$n -> p + e^- + \overline{v_e}$$

and nuclei decay by beta decay in processes like

$$p^{32} \ -> \ S^{32} + e^- + \overline{v_e}$$

is thought to be the result of a more fundamental quark process

$$d \ -> \ u + e^- + \overline{v_e}$$

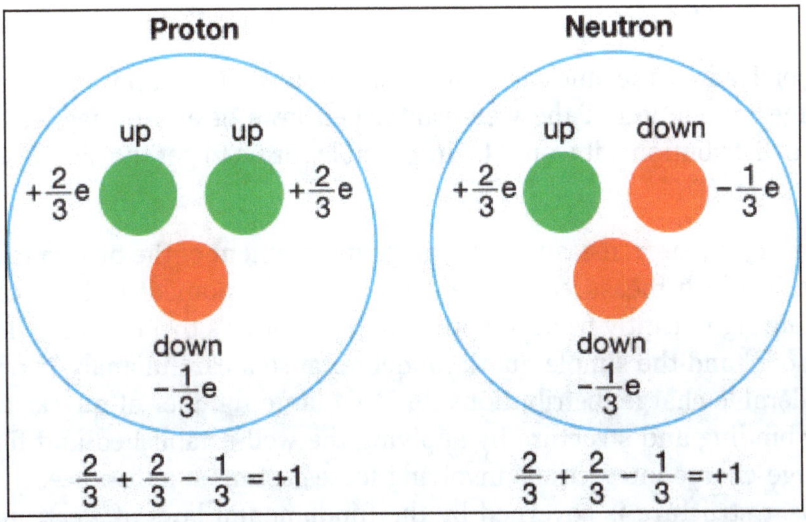

The up quark (u) is the lightest of all quarks, a type of elementary particle, and a major constituent of matter. It, along with the down quark, forms the proton and the neutron. It is part of the first generation of matter, has an electric charge of +2e/3 . Like all quarks, the up quark is an elementary fermion with spin S = ½. In 1964, Gell-Mann and Zweig (independently of each other) proposed the quark model, then consisting only of up, down, and strange quarks. However, while the quark model explained the Eightfold Way, no direct evidence of the existence of quarks was found until 1968 at the Stanford Linear Accelerator Center. Deep inelastic scattering experiments indicated that protons had substructure, and that protons made of more-fundamental particles explained the data (thus confirming the quark model).

The down quark (d) is the second-lightest of all quarks, a type of elementary particle, and a major constituent of matter. Together with the up quark, it forms the neutrons and protons of atomic nuclei. It is part of the first generation of matter, has an electric charge of -e/3 . Like all quarks, the down quark is an elementary fermion with spin S = 1⁄2. Also the peripheral velocity of spinning quarks is greater than the speed of light and under this condition the magnetic forces are stronger than the electric forces which contribute to the very strong QUARK-QUARK INTERACTION. Note that such electromagnetic interactions of natural laws invalidate the theory of the quantum chromodynamics (1973).

Especially in 1964 Gell-Mann after a taxonomy of particles suggested that both protons (p) and neutrons (n) consist of (uud) quarks and (dud) quarks respectively having fractional charges as,

u = +2e/3 and d = -e/3. That is, uud = +e and dud = 0.

Of course such structures imply small charge distributions as,

proton = (+Q = +4e/3, -q = -e/3) and neutron = (+Q = +2e/3, -Q = -2e/3)

which cannot lead to the nuclear structure. Actually, if we apply the fundamental charge-charge interaction of the well-established laws of electromagnetism on such small charge distributions, it would be impossible for us to get the simplest p-n structure of deuterium (D).

Meanwhile in 1933, Stern measured the magnetic moment of the proton to be 2.79 μN and in 1940 F. Bloch measured the neutron magnetic moment to be -1.91 μN. Such results deviate significantly from the predictions of Dirac's theory and invalidate both Yukawa's model and the simple quark model because a careful analysis of them provides considerable charge distributions due to a large number of quarks able to give the nuclear binding and structure by applying the well-established and fundamental laws of charge-charge interactions involving forces acting at a distance. In a paper of 2003 "Nuclear structure is governed by the fundamental laws of electromagnetism" (presented also at the 12th Symposium of the Hellenic Nuclear Physics Society) it is described that the charge distributions of protons and neutrons respectively by a careful analysis of the magnetic moments of nucleons and the deep inelastic scattering experiments. For example for the proton (p) the magnetic moment μ is given by,

μ/S = 2.793(e /M)

where S is the spin of proton, e the net charge of proton and M the mass of proton. Here we see that the above experimental relation cannot be consistent with the simple quark model with the scheme (uud) even in case in which the charge +Q = +4e/3 = 2u is along the periphery and the charge -q = -e/3 = 1d is in the center (deep inelastic scattering experiment). Clearly applying the electromagnetic laws for μ, and the laws of a rotating oblate spheroid (like the proton) we may write for μ and for the spin S (angular momentum) respectively as,

$$\mu = i\pi R^2 = (Q\nu)\pi R^2 \text{ and } S = t\, M\omega R^2 = tM(2\pi\nu)R^2$$

where t is a factor between a rotating sphere [$S = (0.4)M\omega R^2$] and a disc [$S = (0.5) M\omega R^2$]. That is $0.4 < t < 0.5$. Therefore $\mu / S = Q/2t = (2.793)e$. That is, for t = 0.47742 (oblate spheroid) we get for the proton along the periphery +Q = +8e/3 = 4u and in the center -q = -5e/3 = 5d. In the same way for the neutron we get - Q = -8e/3 = 8d along the periphery, and +Q = +8e/3 = 4u in the center. Surprisingly applications of electromagnetic laws on such experimental charge distributions which give for proton extra (4u,5d) charged quarks and for the neutron extra (8d,4u) charged quarks lead exactly to the simplest nuclear binding (-2.2246 MeV) of the deuterium. Moreover such extra charged quarks led to the discovery of 288 quarks in nucleons. As a result the proton has 93 (dud) neutral quark triads. Among them there are 4u charged quarks distributed along the periphery and 5d charged quarks limited in the center. Whereas the neutron has 92 (dud) neutral quark triads and among them are distributed 8d charged quarks along the periphery and 4u charged quarks limited in the center So, the new structure of protons and neutrons is given by,

PROTON = [93(dud) + 4u +5d]. NEUTRON = [92(dud) + 8d + 4u]

After a careful analysis it was found that the masses of down quark (Md) and of up quark (Mu) as,

Md = 3.69348645 MeV/c^2 = mass of 7.23 electrons and Mu = 2.40016645 MeV/c/^{2} = mass of 4.7 electrons

They give the total masses of nucleons as,

Mass of neutron Mn = 939.565378 MeV/c^2 = mass of 1838.68 electrons

Mass of proton Mp = 938.272046 MeV/^{c2} = mass of 1836.15 electrons

Then the difference Mn - Mp = 1.293332 MeV/c^2 = mass of 2.53 electrons is exactly equal to Md-Mu. That is, Md - Mu = 7.23- 4.70 = mass of 2.53 electrons. For example n-p = d-u given by the following equation:

n - p = [92(dud) + 8d + 4u] - [93(dud) + 4u +5d] = 3d -(dud) = d-u = mass of 2.53 electrons.

Note that according to the law of energy and mass given by

$$\Delta w / \Delta m = \Delta E / \Delta M = c2$$

the increase of the electron energy ΔE = 1.293 MeV and the increase of the electron mass ΔM = mass of 2.53 electrons of the one emitting electron in the beta decay are due not to the relative motion of the emitting electron with respect to a falacious ethe or to an observer of special relativity but to the real absorption of the released energy Δw = 1.293 MeV and the mass defect Δm = mass of 2.53 electrons. (Experiments reject relativity).

Antineutrino Absorption and Emission

Here the antineutrino (ν^-) behaves like a neutron because it has negative charge along the periphery and positive one at the center. So it interacts electromagnetically with the up quark (u) having a positive charge (+2e/3) like a photon which as an electric dipole interacting with the charge (-e) of an electron in the photoelectric effect under a weak electromagnetic interaction. Therefore the antineutrino-quark interaction invalidates the so-called Electroweak Theory. In this case the absorption of the antineutrino by the proton gives a neutron and a positron under the two conservation laws of mass and energy as,

$$\nu^- + p = n + e^+$$

$$or \quad \nu^- + \left[93(dud) + 4u + 5d\right] = [\ \left(92(dud) + 4u + 8d\ \right] + e^+$$

$$or \quad \nu^- + (dud) = 3d + e^+$$

$$or \quad \nu^- + u = d + e^+$$

According to the experiments the antineutrino in this case is an energetic particle with a mass equal to 1.8 MeV/c². Since the mass of the positron is equal to 0.51 MeV/c² and using my discovered values of the up and down quarks we can write the conservation law of mass in terms of MeV/c² as,

$$1.8 + 2.4 = 3.69 + 0.51$$

In the same way we can write the conservation law of mass in the antineutrino emission as,

$$n = p + e^- + \nu^-$$

$$or \ \left[(92(dud) + 4u + 8d\ \right] = \left[93(dud) + 4u + 5d\right] + e^- + \nu^-$$

$$or \ d = u + e^- + \nu^-$$

Since here the mass of the emitting energetic antineutrino is equal to 0.78 MeV/c² we can write the conservation law of mass in terms of MeV/c² as,

$$3.69 = 2.4 + 0.51 + 0.78$$

Strange Quark

The strange quark or s quark (from its symbol, s) is the third lightest of all quarks, a type of elementary particle. Strange quarks are found in subatomic particles called hadrons. Example of hadrons containing strange quarks include kaons (K), strange D mesons (D_s), Sigma baryons (Σ), and other strange particles.

According to the IUPAP the symbol s is the official name, while strange is to be considered only as a mnemonic. The name sideways has also been used because the s quark has a I_3 value of 0 while the u ("up") and d ("down") quarks have values of $+\frac{1}{2}$ and $-\frac{1}{2}$ respectively.

Along with the charm quark, it is part of the second generation of matter, and has an electric charge of $\frac{1}{3}e$ and a bare mass of 95^{+5}_{-5} MeV/c². Like all quarks, the strange quark is an elementaryfermion with spin $\frac{1}{2}$, and experiences all four fundamental interactions: gravitation, electromagnetism, weak interactions, and strong interactions. The antiparticle of the strange quark is the strange antiquark (sometimes called *antistrange quark* or simply *antistrange*), which differs from it only in that some of its properties have equal magnitude but opposite sign.

The first strange particle (a particle containing a strange quark) was discovered in 1947 (kaons), but the existence of the strange quark itself (and that of the up and down quarks) was only postulated in 1964 by Murray Gell-Mann and George Zweig to explain the Eightfold Way classification scheme of hadrons. The first evidence for the existence of quarks came in 1968, in deep inelastic scattering experiments at the Stanford Linear Accelerator Center. These experiments confirmed the existence of up and down quarks, and by extension, strange quarks, as they were required to explain the Eightfold Way.

Charm Quark

The charm quark is one of six quarks that, along with leptons, form the basic building blocks of ordinary matter. It is hundreds of times more massive than the up and down quarks that make up protons and neutrons.

Theorists had predicted the existence of the charm quark to explain the absence of an expected particle interaction. When the charm quark did turn up, it was as a constituent of the J/psi particle, whose discovery in 1974 finally convinced physicists that quarks were real.

Particles containing a charm quark are known as either “charmed particles” or “charmonia.” They have only a fleeting existence before decaying into more conventional particles. Many experiments have studied their properties. At facilities known as “charm factories,” large numbers of charm-containing particles can be produced with little contamination from other types of particles. An electron-positron collider in Beijing, for example, is expected to ultimately produce 10 billion J/psi decays in one year’s running time. This lcollevel of production allows scientists to observe subatomic processes with great precision, and may reveal subtle signs of new physics phenomena predicted by theorists.

Hadron

Hadron, any member of a class of subatomic particles that are built from quarks and thus react through the agency of the strong force. The hadrons embrace mesons, baryons (e.g., protons, neutrons, and sigma particles), and their many resonances. All observed subatomic particles are hadrons except for the gauge bosons of the fundamental interactions and the leptons. Except for protons and neutrons that are bound in atomic nuclei, all hadrons have short lives and are produced in the high-energy collisions of subatomic particles. The other three basic forces of nature also affect hadron behaviour: all hadrons are subject to gravitation; charged hadrons obey electromagnetic laws; and some hadrons break up by way of the weak force (as in radioactive decay), while others decay via the strong and the electromagnetic forces.

Properties

According to the quark model, the properties of hadrons are primarily determined by their so-called *valence quarks*. For example, a proton is composed of two up quarks (each with electric charge $+\frac{2}{3}$, for a total of $+\frac{4}{3}$ together) and one down quark (with electric charge $-\frac{1}{3}$). Adding these together yields the proton charge of +1. Although quarks also carry color charge, hadrons must have zero total color charge because of a phenomenon called color confinement. That is, hadrons must be "colorless" or "white". These are the simplest of the two ways: three quarks of different colors, or a quark of one color and an antiquarkcarrying the corresponding anticolor. Hadrons with the first arrangement are called baryons, and those with the second arrangement are mesons.

Hadrons, however, are not composed of just three or two quarks, because of the strength of the strong force. More accurately, strong force gluons have enough energy (E) to have resonances composed of massive (m) quarks ($E > mc^2$) . Thus, virtual quarks and antiquarks, in a 1:1 ratio, form the majority of massive particles inside a hadron. The two or three quarks that compose a hadron are the excess of quarks vs. antiquarks, and so too in the case of anti-hadrons (anti-particles). Because the virtual quarks are not stable wave packets (quanta), but an irregular and transient phenomenon, it is not meaningful to ask which quark is real and which virtual; only the excess is apparent from the outside in the form of a hadron. Massless virtual gluons compose the numerical majority of particles inside hadrons.

Like all subatomic particles, hadrons are assigned quantum numbers corresponding to the representations of the Poincaré group: $J^{PC}(m)$, where J is the spin quantum number, P the intrinsic parity (or P-parity), C the charge conjugation (or C-parity), and $m$the particle's mass. Note that the mass of a hadron has very little to do with the mass of its valence quarks; rather, due to mass–energy equivalence, most of the mass comes from the large amount of energy associated with the strong interaction. Hadrons may also carry flavor quantum numbers such as isospin (G parity), and strangeness. All quarks carry an additive, conserved quantum number called a baryon number (B), which is $+\frac{1}{3}$ for quarks and $-\frac{1}{3}$ for antiquarks. This means that baryons (groups of three quarks) have $B = 1$ whereas mesons have $B = 0$.

Hadrons have excited states known as resonances. Each ground state hadron may have several excited states; several hundreds of resonances have been observed in experiments. Resonances decay extremely quickly (within about 10^{-24} seconds) via the strong nuclear force.

In other phases of matter the hadrons may disappear. For example, at very high temperature and high pressure, unless there are sufficiently many flavors of quarks, the theory of quantum chromodynamics (QCD) predicts that quarks and gluons will no longer be confined within hadrons, "because the strength of the strong interaction diminishes with energy". This property, which is known as asymptotic freedom, has been experimentally confirmed in the energy range between 1 GeV (gigaelectronvolt) and 1 TeV (teraelectronvolt).

All free hadrons except the proton and antiproton are unstable.

Baryon

Baryon, any member of one of two classes of hadrons (particles built from quarks and thus experiencing the strong nuclear force). Baryons are heavy subatomic particles that are made up of three quarks. Both protons and neutrons, as well as other particles, are baryons. (The other class of hadronic particle is built from a quark and an antiquark and is called a meson.)

Baryons are characterized by a baryon number, *B*, of 1. Their antiparticles, called antibaryons, have a baryon number of −1. An atom containing, for example, one proton and one neutron (each with a baryon number of 1) has a baryon number of 2. In addition to their differences in composition, baryons and mesons can be distinguished from one another by spin: the three quarks that make up a baryon can only produce half-integer values, while meson spins always add up to integer values.

Baryons are strongly interacting fermions; that is, they are acted on by the strong nuclear force and are described by Fermi–Dirac statistics, which apply to all particles obeying the Pauli exclusion principle. This is in contrast to the bosons, which do not obey the exclusion principle.

Baryons, along with mesons, are hadrons, particles composed of quarks. Quarks have baryon numbers of $B = \frac{1}{3}$ and antiquarks have baryon numbers of $B = -\frac{1}{3}$. The term "baryon" usually refers to *triquarks*—baryons made of three quarks ($B = \frac{1}{3} + \frac{1}{3} + \frac{1}{3} = 1$).

Other exotic baryons have been proposed, such as pentaquarks—baryons made of four quarks and one antiquark ($B = \frac{1}{3} + \frac{1}{3} + \frac{1}{3} + \frac{1}{3} - \frac{1}{3} = 1$), but their existence is not generally accepted. The particle physics community as a whole did not view their existence as likely in 2006, and in 2008, considered evidence to be overwhelmingly against the existence of the reported pentaquarks. However, in July 2015, the LHCb experiment observed two resonances consistent with pentaquark states in the $\Lambda_b^0 \rightarrow J/\psi K^- p$ decay, with a combined statistical significance of 15σ.

In theory, heptaquarks (5 quarks, 2 antiquarks), nonaquarks (6 quarks, 3 antiquarks), etc. could also exist.

Baryonic Matter

Nearly all matter that may be encountered or experienced in everyday life is baryonic matter, which includes atoms of any sort, and provides those with the property of mass. Non-baryonic matter, as implied by the name, is any sort of matter that is not composed primarily of baryons. This might include neutrinos and free electrons, dark matter, supersymmetric particles, axions, and black holes.

The very existence of baryons is also a significant issue in cosmology because it is assumed that the Big Bang produced a state with equal amounts of baryons and antibaryons. The process by which baryons came to outnumber their antiparticles is called baryogenesis.

Baryogenesis

Experiments are consistent with the number of quarks in the universe being a constant and, to be more specific, the number of baryons being a constant; in technical language, the total baryon number appears to be conserved. Within the prevailing Standard Model of particle physics, the number of baryons may change in multiples of three due to the action of sphalerons, although this is rare and has not been observed under experiment. Some grand unified theories of particle physics also predict that a single proton can decay, changing the baryon number by one; however, this has not yet been observed under experiment. The excess of baryons over antibaryons in the present universe is thought to be due to non-conservation of baryon number in the very early universe, though this is not well understood.

Properties

Isospin and Charge

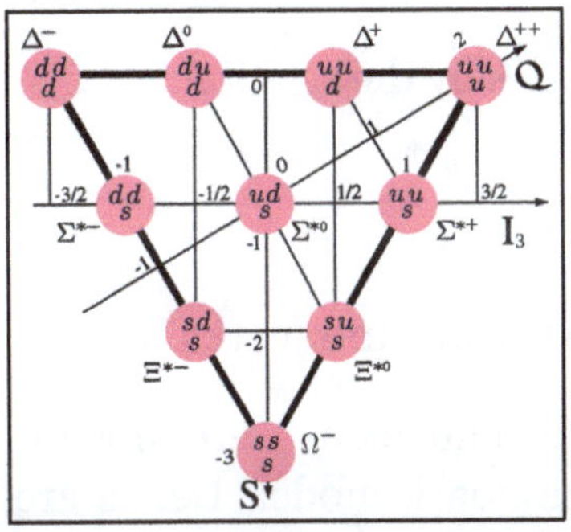

Combinations of three u, d or squarks forming baryons with a spin-$\frac{3}{2}$ form the *uds baryon decuplet*.

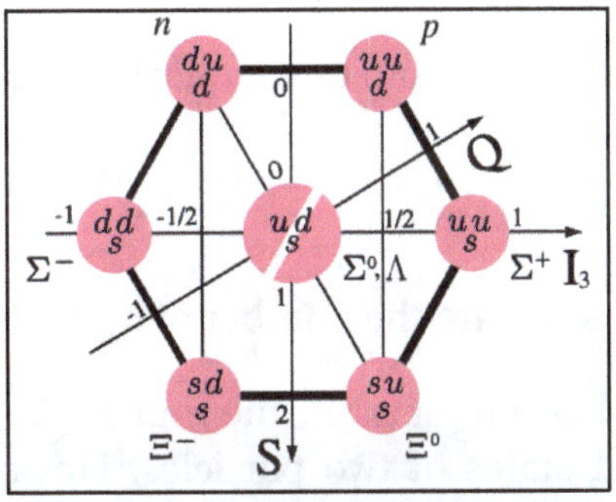

Combinations of three u, d or squarks forming baryons with a spin-$\frac{1}{2}$ form the *uds baryon octet*.

The concept of isospin was first proposed by Werner Heisenberg in 1932 to explain the similarities between protons and neutrons under the strong interaction. Although they had different electric charges, their masses were so similar that physicists believed they were the same particle. The different electric charges were explained as being the result of some unknown excitation similar to spin. This unknown excitation was later dubbed *isospin* by Eugene Wigner in 1937.

This belief lasted until Murray Gell-Mann proposed the quark model in 1964 (containing originally only the u, d, and s quarks). The success of the isospin model is now understood to be the result of the similar masses of u and d quarks. Since u and d quarks have similar masses, particles made of the same number then also have similar masses. The exact specific u and d quark composition determines the charge, as u quarks carry charge $+\frac{2}{3}$ while d quarks carry charge $-\frac{1}{3}$. For example, the four Deltas all have different charges (Δ^{++} (uuu), Δ^{+} (uud), Δ^{0} (udd), Δ^{-} (ddd)), but have similar masses (~1,232 MeV/c^2) as they are each made of a combination of three u and d quarks. Under the isospin model, they were considered to be a single particle in different charged states.

The mathematics of isospin was modeled after that of spin. Isospin projections varied in increments of 1 just like those of spin, and to each projection was associated a "charged state". Since the "Delta particle" had four "charged states", it was said to be of isospin $I = \frac{3}{2}$. Its "charged states" Δ^{++}, Δ^{+}, Δ^{0}, and Δ^{-}, corresponded to the isospin projections $I_3 = +\frac{3}{2}$, $I_3 = +\frac{1}{2}$, $I_3 = -\frac{1}{2}$, and $I_3 = -\frac{3}{2}$, respectively. Another example is the "nucleon particle". As there were two nucleon "charged states", it was said to be of isospin $\frac{1}{2}$. The positive nucleon N^{+} (proton) was identified with $I_3 = +1/2$ and the neutral nucleon N^{0} (neutron) with $I_3 = -\frac{1}{2}$. It was later noted that the isospin projections were related to the up and down quark content of particles by the relation:

$$I_3 = \frac{1}{2}[(n_{\mathrm{u}} - n_{\bar{\mathrm{u}}}) - (n_{\mathrm{d}} - n_{\bar{\mathrm{d}}})],$$

where the n's are the number of up and down quarks and antiquarks.

In the "isospin picture", the four Deltas and the two nucleons were thought to be the different states of two particles. However, in the quark model, Deltas are different states of nucleons (the N^{++} or N^{-} are forbidden by Pauli's exclusion principle). Isospin, although conveying an inaccurate picture of things, is still used to classify baryons, leading to unnatural and often confusing nomenclature.

Flavour Quantum Numbers

The strangeness flavour quantum number S was noticed to go up and down along with particle mass. The higher the mass, the lower the strangeness (the more s quarks). Particles could be described with isospin projections (related to charge) and strangeness (mass). As other quarks were discovered, new quantum numbers were made to have similar description of udc and udb octets and decuplets. Since only the u and d mass are similar, this description of particle mass and charge in terms of isospin and flavour quantum numbers works well only for octet and decuplet made of one u, one d, and one other quark, and breaks down for the other octets and decuplets (for example, ucb octet and decuplet). If the quarks all had the same mass, their behaviour would be called *symmetric*, as they would all behave in the same way to the strong interaction. Since quarks do not have the same mass, they do not interact in the same way (exactly like an electron placed in an electric field will accelerate more than a proton placed in the same field because of its lighter mass), and the symmetry is said to be broken.

It was noted that charge (Q) was related to the isospin projection (I_3), the baryon number (B) and flavour quantum numbers (S, C, B', T) by the Gell-Mann–Nishijima formula:

$$Q = I_3 + \frac{1}{2}(B + S + C + B' + T),$$

where S, C, B', and T represent the strangeness, charm, bottomness and topnessflavour quantum numbers, respectively. They are related to the number of strange, charm, bottom, and top quarks and antiquark according to the relations:

$$S = -\left(n_s - n_{\bar{s}}\right)$$

$$C = +\left(n_c - n_{\bar{c}}\right),$$

$$B' = -(n_b - n_{\bar{b}}),$$

$$T = +(n_t - n_{\bar{t}}),$$

meaning that the Gell-Mann–Nishijima formula is equivalent to the expression of charge in terms of quark content:

$$Q\frac{2}{3}\left[\left(n_u - n_{\bar{u}}\right) + \left(n_c - n_{\bar{c}}\right) + \left(n_t - n_{\bar{t}}\right)\right] - \frac{1}{3}\left[\left(n_d - n_{\bar{d}}\right) + \left(n_s - n_{\bar{s}}\right) + \left(n_b - n_{\bar{b}}\right)\right].$$

Spin, Orbital Angular Momentum and Total Angular Momentum

Spin (quantum number S) is a vector quantity that represents the "intrinsic" angular momentum of a particle. It comes in increments of $\frac{1}{2}$ ħ (pronounced "h-bar"). The ħ is often dropped because it is the "fundamental" unit of spin, and it is implied that "spin 1" means "spin 1 ħ". In some systems of natural units, ħ is chosen to be 1, and therefore does not appear anywhere.

Quarks are fermionic particles of spin $\frac{1}{2}$ ($S = \frac{1}{2}$). Because spin projections vary in increments of 1 (that is 1 ħ), a single quark has a spin vector of length $\frac{1}{2}$, and has two spin projections ($S_z = +\frac{1}{2}$ and $S_z = -\frac{1}{2}$). Two quarks can have their spins aligned, in which case the two spin vectors add to make a vector of length $S = 1$ and three spin projections

($S_z = +1$, $S_z = 0$, and $S_z = -1$). If two quarks have unaligned spins, the spin vectors add up to make a vector of length $S = 0$ and has only one spin projection ($S_z = 0$), etc. Since baryons are made of three quarks, their spin vectors can add to make a vector of length $S = \frac{3}{2}$, which has four spin projections ($S_z = +\frac{3}{2}$, $S_z = +\frac{1}{2}$, $S_z = -\frac{1}{2}$, and $S_z = -\frac{3}{2}$), or a vector of length $S = \frac{1}{2}$ with two spin projections ($S_z = +\frac{1}{2}$, and $S_z = -\frac{1}{2}$).

There is another quantity of angular momentum, called the orbital angular momentum(azimuthal quantum number L), that comes in increments of 1 ħ, which represent the angular moment due to quarks orbiting around each other. The total angular momentum(total angular momentum quantum number J) of a particle is therefore the combination of intrinsic angular momentum (spin) and orbital angular momentum. It can take any value from $J = |L - S|$ to $J = |L + S|$, in increments of 1.

Baryon angular momentum quantum numbers for L = 0, 1, 2, 3				
Spin (S)	**Orbital angular momentum (L)**	**Total angular momentum (J)**	**Parity (P)**	**Condensed notation (J^P)**
$\frac{1}{2}$	0	$\frac{1}{2}$	+	$\frac{1}{2}^+$
	1	$\frac{3}{2}, \frac{1}{2}$	−	$\frac{3}{2}^-, \frac{1}{2}^-$
	2	$\frac{5}{2}, \frac{3}{2}$	+	$\frac{5}{2}^+, \frac{3}{2}^+$
	3	$\frac{7}{2}, \frac{5}{2}$	−	$\frac{7}{2}^-, \frac{5}{2}^-$
$\frac{3}{2}$	0	$\frac{3}{2}$	+	$\frac{3}{2}^+$
	1	$\frac{5}{2}, \frac{3}{2}, \frac{1}{2}$	−	$\frac{5}{2}^-, \frac{3}{2}^-, \frac{1}{2}^-$
	2	$\frac{7}{2}, \frac{5}{2}, \frac{3}{2}, \frac{1}{2}$	+	$\frac{7}{2}^+, \frac{5}{2}^+, \frac{3}{2}^+, \frac{1}{2}^+$
	3	$\frac{9}{2}, \frac{7}{2}, -, \frac{3}{2}$	−	$\frac{9}{2}^-, \frac{7}{2}^-, \frac{5}{2}^-, \frac{3}{2}^-$

Particle physicists are most interested in baryons with no orbital angular momentum ($L = 0$), as they correspond to ground states—states of minimal energy. Therefore, the two groups of baryons most studied are the $S = \frac{1}{2}$; $L = 0$ and $S = \frac{3}{2}$; $L = 0$, which corresponds to $J = \frac{1}{2}^+$ and $J = \frac{3}{2}^+$, respectively, although they are not the only ones. It is also possible to obtain $J = \frac{3}{2}^+$ particles from $S = \frac{1}{2}$ and $L = 2$, as well as $S = \frac{3}{2}$ and $L = 2$. This phenomenon of having multiple particles in the same total angular momentum configuration is called *degeneracy*. How to distinguish between these degenerate baryons is an active area of research in baryon spectroscopy.

Parity

If the universe were reflected in a mirror, most of the laws of physics would be identical—things would behave the same way regardless of what we call "left" and what we call "right". This concept of mirror reflection is called intrinsic parity" or simply "parity" (*P*). Gravity, the electromagnetic force, and the strong interaction all behave in the same way regardless of whether or not the universe is reflected in a mirror, and thus are said to conserve parity (P-symmetry). However, the weak interaction does distinguish "left" from "right", a phenomenon called parity violation (P-violation).

Based on this, if the wavefunction for each particle (in more precise terms, the quantum field for each particle type) were simultaneously mirror-reversed, then the new set of wavefunctions would perfectly satisfy the laws of physics (apart from the weak interaction). It turns out that this is not quite true: for the equations to be satisfied, the wavefunctions of certain types of particles have to be multiplied by −1, in addition to being mirror-reversed. Such particle types are said to have negative or odd parity ($P = -1$, or alternatively $P = -$), while the other particles are said to have positive or even parity ($P = +1$, or alternatively $P = +$).

For baryons, the parity is related to the orbital angular momentum by the relation:

$$P = (-1)^L.$$

As a consequence, baryons with no orbital angular momentum ($L = 0$) all have even parity ($P = +$).

Nomenclature

Baryons are classified into groups according to their isospin (*I*) values and quark (*q*) content. There are six groups of baryons—nucleon (N), Delta (Δ), Lambda (Λ), Sigma (Σ), Xi (Ξ), and Omega (Ω). The rules for classification are defined by the Particle Data Group. These rules consider the up (u), down (d) and strange (s) quarks to be *light* and the charm (c), bottom (b), and top (t) quarks to be *heavy*. The rules cover all the particles that can be made from three of each of the six quarks, even though baryons made of top quarks are not expected to exist because of the top quark's short lifetime. The rules do not cover pentaquarks.

- Baryons with three $u\, and\, /\, or\, d\, quarks\ are\, N's\ (I=\frac{1}{2})\ or\, \Delta's\ (I=\frac{3}{2})$.
- Baryons with two u and/or d quarks are $\Lambda's\ (I=0)\ or\, \Sigma's\ (I=1)$. If the third quark is heavy, its identity is given by a subscript.
- Baryons with one u or d quark are $\Xi's\ (I=\frac{1}{2})$. One or two subscripts are used if one or both of the remaining quarks are heavy.
- Baryons with no u or d quarks are $\Omega's\ (I=0)$, and subscripts indicate any heavy quark content.
- Baryons that decay strongly have their masses as part of their names. For example, Σ^o does not decay strongly, but $\Delta^{++}(1232)$ does.

It is also a widespread (but not universal) practice to follow some additional rules when distinguishing between some states that would otherwise have the same symbol.

- Baryons in total angular momentum $J = \frac{3}{2}$ configuration that have the same symbols as their $J = \frac{1}{2}$ counterparts are denoted by an asterisk (*).
- Two baryons can be made of three different quarks in $J = \frac{1}{2}$ configuration. In this case, a prime (′) is used to distinguish between them.
- *Exception*: When two of the three quarks are one up and one down quark, one baryon is dubbed Λ while the other is dubbed Σ.

Quarks carry a charge, so knowing the charge of a particle indirectly gives the quark content. For example, the rules above say that a Λ^+_c contains a c quark and some combination of two u and/or d quarks. The c quark has a charge of ($Q = +\frac{2}{3}$), therefore the other two must be a u quark ($Q = +\frac{2}{3}$), and a d quark ($Q = -\frac{1}{3}$) to have the correct total charge ($Q = +1$).

Table of Baryons

Particle	Symbol	Makeup	Rest mass MeV/c²	Spin	B	S	Lifetime (seconds>	Decay Modes
Proton	p	uud	938.3	1/2	+1	0	Stable	...
Neutron	n	ddu	939.6	1/2	+1	0	920	$pe^-\bar{v}_e$
Lambda	Λ^o	uds	1115.6	1/2	+1	-1	2.6 x10^{-10}	$p\pi^-$, $n\pi^o$
Sigma	Σ^+	uus	1189.4	1/2	+1	-1	0.8 x10^{-10}	$p\pi^o$, $n\pi^+$
Sigma	Σ^o	uds	1192.5	1/2	+1	-1	6x10^{-20}	$\Lambda^o\gamma$
Sigma	Σ^-	dds	1197.3	1/2	+1	-1	1.5 x10^{-10}	$n\pi^-$

Delta	Δ^{++}	uuu	1232	3/2	+1	0	0.6 x10^{-23}	$p\pi^{+}$
Delta	Δ^{+}	uud	1232	3/2	+1	0	0.6 x10^{-23}	$p\pi^{0}$
Delta	Δ^{0}	udd	1232	3/2	+1	0	0.6 x10^{-23}	$n\pi^{0}$
Delta	Δ^{-}	ddd	1232	3/2	+1	0	0.6 x10^{-23}	$n\pi^{-}$
Xi Cascade	Ξ^{0}	uss	1315	1/2	+1	-2	2.9 x10^{-10}	$\Lambda^{0}\pi^{0}$
Xi Cascade	Ξ^{-}	dss	1321	1/2	+1	-2	1.64 x10^{-10}	$\Lambda^{0}\pi^{-}$
Omega	Ω^{-}	sss	1672	3/2	+1	-3	0.82 x10^{-10}	$\Xi^{0}\pi^{-}$, $\Lambda^{0}K^{-}$
Lambda	Λ^{+}_{c}	udc	2281	1/2	+1	0	2x10^{-13}	...

As an example of the notation in the column of decay modes, the sigma + decay mode is written,

$$p\pi^{0},\ n\pi^{+}$$

which implies there are two common decay modes,

$$\Sigma^{+} \rightarrow p + \pi^{0} \quad and \quad \Sigma^{+} \rightarrow n + \pi^{+}$$

Particles which decay by the strong interaction typically do so on a time scale of about 10-23 seconds, and some of the baryons listed in the table above have lifetimes on the order of 10-10 seconds. The reason for those longer lifetimes is that they are forbidden by some kind of conservation law from decaying by the strong interaction, and must decay by the weak interaction. These decays involve quark transformations which can be represented by Feynman diagrams.There are dozens of other baryons (i.e., combinations of three quarks), but most of them decay by the strong interaction and are so short lived that they don't leave a measurable track in a bubble chamber or other detection device. These baryons have been detected as "resonances" or peaks in the probability for particle interactions.

Nucleon

The atom is electrically neutral due to equal balance of negative charged particles electrons and positively charged particles. And the neutrons are neutral particles. The negative particles are revolved around a nucleus and nucleus contains positive particles protons and neutral neutrons. The total number of neutrons and protons is called Nucleons and this is the mass number while the number of only protons is called atomic number of elements.

Nucleons are held up together inside the nucleus due to nuclear force. Nucleons are further made up of three quarks. These quarks are help up together due to strong interactions.

Proton and neutron can also be found alone, not a part of another nucleus. A proton is the nucleus of H – 1 atom or hydrogen atom. Neutrons can be found in nuclear reactions. A neutron is unstable in nature. As said above nucleon is made of three quarks. They can be up or down quark. A proton consists of one down quark and two up quarks. A neutron consists of one up quark and two down quarks. The strong interaction between the quarks is carried by a particle called gluon.

Charge of up quark is $\frac{+2e}{3}$ and that of down quark is $\frac{-e}{3}$. Hence the charge of proton is +e and that of neutron is 0.. Therefore neutron is neutral in nature. Mass of neutron and proton is almost equal. Mass of neutron is 1.6749×10^{-27} kg and that of proton is 1.6726×10^{-27} kg.

Proton and neutron together forms an isospin doublet. This is why their masses is approximately equal. Neutron is 0.1 percent heavier than proton. Proton and neutron both have same spin i.e. $\frac{1}{2}$. Therefore they lie in the category of fermions and hence Pauli's exclusion principle is applicable on them. According to this they cannot lie in the same quantum state. Therefore they spread out in the nuclear shells similar to electron shells.

A neutron is unstable and hence undergoes beta ($\beta-$) decay and converts into proton releasing an electron and electron antineutrino. While a proton is stable in nature and has larger half life. Inside a nucleus proton and neutron may or may not undergo decay process depending on the nuclide.

Both protons and neutrons have corresponding antiparticles called as antiproton and the antineutron respectively. These anti particles have the same mass but opposite charge as the proton and neutron respectively. They interact in the same way as the nucleons. In particular, anti nucleons can bind to form an "anti nucleus". Till now, scientists have created anti deuterium and anti helium nuclei.

Nucleon Number

Nucleon number is also known as mass number. Nucleon number of an atom is given by the sum of total number of protons and neutrons or total number of nucleons present in the nucleus of that atom.

For each different isotope of an element, mass number is different. Mass number is different from the atomic number. Atomic number is the total number of protons present in the nucleus of an atom. Therefore total number of neutrons can be found out by the difference of mass number and atomic number.

For an atom:

$$\text{Mass number}\ (A) = \text{atomic number}\ (Z) + \text{number of neutrons}\ (N)$$

In a radioactive decay, mass number of an atom changes. Different types of radioactive decays are characterized by the change in atomic number and nucleon number. This characterization is done according to *Fajans and Soddy's law of radioactive displacement.*

For e.g. :

- U-238 undergo radioactive decay by alpha decay, where the number of protons and neutrons both decreases by two in the form of an alpha particle. Therefore the number of neutrons and atomic number decreases by 2 from 146 to 144 and from 92 to 90 respectively, hence the nucleon number decreases by 4 from 238 to 234. This results in an atom of thorium-234 and an alpha particle .
- On the other side, carbon-14 undergoes decays naturally by beta decay. Here one neutron is transformed into a proton by emitting an electron and anti particle of neutron (anti - neutrino). Therefore the atomic number increases by 1 from 6 to 7 and the nucleon or mass number remains the same i.e. 14, whereas the number of neutrons decreases by 1 from 8 to 7. This results in the formation of nitrogen-14 atom, with 7 protons and neutrons.

The nucleon number gives helps in estimation of the isotopic mass measured in atomic mass unit (u). For C – 12, the isotopic mass is exactly 12 because the atomic mass unit is 1/12 of the mass of C- 12 . The isotopic mass for other isotopes is normally within 0.1 u of the mass number. For example, nucleon number of Cl – 35 is 35 and its isotopic mass is equal to 34.96885. The difference between mass number and isotopic mass which is known as the Mass Defect is mainly due to two reasons:

- The neutron inside atoms is little heavier than the proton which increases the mass of a nucleus with more number of neutrons than protons relative to the atomic mass unit scale based on C - 12 with equal numbers of protons and neutrons.
- Second reason is that the nuclear binding energy varies from nucleus to nucleus. A nucleus with higher binding energy has a lesser total energy resulting in a lower mass according to Einstein's mass-energy relation $E = mc^2$. For Cl – 35, the isotopic mass is less than 35, therefore it must be the dominant factor.

Mass of Nucleon

A nucleon is of two types: proton and neutron. Protons and neutrons have approximately the same mass, mass of neutron is 1.6749×10^{-27} kg and that of proton is 1.6726×10^{-27} kg. Proton and neutron together forms an isospin doublet. This is why their masses is approximately equal. Neutron is 0.1 percent heavier than proton. So we won't make any difference between proton and neutron. We will refer to them by the composite term "nucleon", because they are found in the nuclei of the atoms. So, if matter is made of small atoms, a question arises that where is the mass located in atoms?

The electrons are very light as compared to nucleons. 1 nucleon weighs as much as two thousand electrons. For all practical purposes, the mass of atoms is considered as the mass of nucleons. It is pondered upon that nucleons consist of 3 visible quarks, which can be considered right. It can be regarded as logical to visualize that mass of each quark is $\frac{1}{3}$ times mass of an individual nucleon, but that›s not right actually. The mass of the 3 quarks together in the nucleons weighs only about 1 - 2 % of the mass of the nucleons, and there is a basic query that where did the rest 89 % come from ?

First, we have to think that nucleons are not static objects with 3 ingredients i.e. quarks. A nucleon is made of 3 very light quarks which are held together by the aid of interaction called strong interaction is is binding. These three quarks move at very high velocities inside the nucleon. To understand this, imagine 3 ping balls in a jackpot machine. These balls are not the most important thing to consider; rather, one should focus on what's forcing them to move. Similarly, think of nucleons as three quark flecks which are tossed rigorously inside a small subatomic tornado. Here tornado is far more important to consider than the tiny flecks.

This can be thought of Einstein's energy mass equation, $E = mc^2$. This equation predicts the mass and energy are equal in nature to each other. talking about the mass of nucleons, we will notice that around 98 percent of the mass of the universe does not exist as the mass in the usual way we think about it but rather the mass exists in the form of energy. The mass of the quarks themselves comes from a different source. The mass of the quarks and leptons was given by the scientist Higgs Boson. However, it's not the dominant source of mass in the universe. The mass we do have is due to the well understood nuclear physics (due to strong nuclear force or strong interaction).

Energy per Nucleon

Conventionally mass of a nucleus is sum of mass all nucleons i.e sum of mass of protons and neutrons present in the nucleus. The mass of a nucleus is less than the summation of mass of the individual nucleons that make up that nucleus. Here comes the concept of nuclear binding energy.

Nuclear binding energy is the energy needed to break the nucleus into its individual nucleons. This can be also expressed as the amount of energy released when the nucleus is formed by combining separate nucleons.

Binding energy is equal to the potential nuclear energy decrement of the nucleons when they combine to form nucleus. This is equivalent to the work done on the nucleons by the strong interaction or strong nuclear force.

Binding energy is the energy that makes the nucleons to be hold together and is associated with the strong interaction.

The mass of a nucleus is less than the summation of mass of the individual nucleons that form that nucleus. The mass difference (ΔM) between the mass of nucleus and total mass of individual nucleons is equivalent to the binding energy of the nucleus.

Einstein has given an equation which shows the relationship between mass difference and binding energy as follows :

$$E_{binding} = \Delta Mc^2$$

(There exist is a factor for conversion of mass (u) to energy in MeV on the AQA board data sheet which saves us converting the mass into kg and then using E = mc^2 to find out the energy in joules and then convert the energy in J to MeV).

ΔM is difference between mass of nucleus and sum of mass of individual nucleons

$$\Delta M = Z \times M_p + (A - Z) \times M_n - M_{nucleus}$$

Where:

ΔM is mass difference

M_p is the mass of a proton

M_n is the mass of a neutron

$M_{nucleus}$ is mass of the formed nucleus

Z is atomic number (number of proton present in the nucleus)

A is nucleon number or mass number (total number of protons and neutrons present in the nucleus)

Binding Energy / Nucleon: We can work out the binding energy per nucleon and the number of nucleons if we know the binding energy in a nucleus. Binding energy per nucleon is the average energy needed to remove each nucleon.

Stability of a nucleus depends on the binding energy per nucleon. Stability of nucleus is directly proportional to the binding energy per nucleon. Stability increases with the increase in binding energy per nucleon.

We can draw a graph of binding energy per nucleon versus nucleon number which gives a smooth curve - with one remarkable anomaly - helium (^{4}He) the binding energy per nucleon is (28.38 / 4) M eV = 7.1 M eV.

This is not what we would expect on the general curve.

Binding Energy Curve

The peak in the binding energy curve in the region of stability near iron is because of the fact that either the breaking up of heavier nuclei by fission or combining of lighter nuclei by fusion will give nuclei which are more tightly bound i.e. less mass per nucleon.

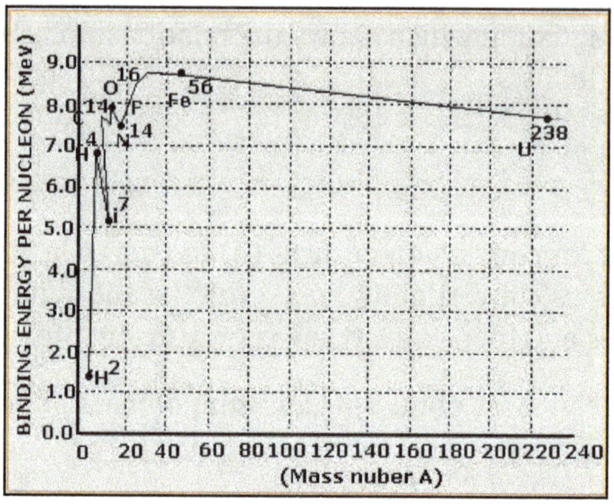

Properties

Quark Composition of a Nucleon

Proton (p): uud Neutron (n): udd Antiproton (p): uud Antineutron (n): udd

A proton (p) is composed of two up quarks (u) and one down quark (d): uud. A neutron (n) has one up quark (u) and two down quarks (d): udd. An antiproton (p) has two up antiquarks (u) and one down antiquark (d): uud. An antineutron (n) has one up antiquark (u) and two down antiquarks (d): udd. The color charge (color assignment) of individual quarks is arbitrary, but all three colors (red, green, blue) must be present.

Protons and neutrons are best known in their role as nucleons, i.e., as the components of atomic nuclei, but they also exist as free particles. Free neutrons are unstable, with a half-life of around 13 minutes, but they are common in nature and have important applications. Singular protons, not bound to other nucleons, are usually regarded as the nuclei of hydrogen atoms or ions, but in some extreme cases (cosmic rays, proton beams), they may be regarded as free protons.

Neither the proton nor neutron is an elementary particle, meaning each is composed of smaller parts, namely three quarks each. A proton is composed of two up quarks and one down quark, while the neutron has one up quark and two down quarks. Quarks are held together by the strong force, or equivalently, by gluons, which mediate the strong force.

An up quark has electric charge $+\frac{2}{3}e$, and a down quark has charge $-\frac{1}{3}e$, so the summed electric charges of proton and neutron are $+e$ and 0, respectively. Thus, the neutron has a charge of 0 (zero), and therefore is eletrically neutral; indeed, the term "neutron" comes from the fact that a neutron is electrically neutral.

The mass of the proton and neutron is quite similar: The proton is 1.6726×10^{-27} kg or 938.27 MeV/c^2, while the neutron is 1.6749×10^{-27} kg or 939.57 MeV/c^2. The neutron is roughly 0.13% heavier. The similarity in mass can be explained roughly by the slight difference in masses of up quarks and down quarks composing the nucleons. However, a detailed explanation remains an unsolved problem in particle physics.

The spin of both protons and neutrons is $\frac{1}{2}$, which means they are fermions and, like electrons (and unlike bosons), are subject to the Pauli exclusion principle, a very important phenomenon in nuclear physics: protons and neutrons in an atomic nucleus cannot all be in the same quantum state; instead they spread out into nuclear shells analogous to electron shells in chemistry. Also important, this spin (of proton and neutron) is the source of nuclear spinin larger nuclei. Nuclear spin is best known for its crucial role in the NMR/MRI technique for chemical and biochemical analyses.

The magnetic moment of a proton, denoted μ_p, is 2.79 nuclear magnetons (μ_N), while the magnetic moment of a neutron is $\mu_n = -1.91\ \mu_N$. These parameters are also important in NMR/MRI.

Stability

A neutron in free state is an unstable particle, with a half-life around ten minutes. It undergoes β^-decay (a type of radioactive decay) by turning into a proton while emitting an electron and an electron antineutrino. A proton by itself is thought to be stable, or at least its lifetime is too long to measure. This is an important discussion in particle physics.

Inside a nucleus, on the other hand, combined protons and neutrons (nucleons) can be stable or unstable depending on the nuclide, or nuclear species. Inside some nuclides, a neutron can turn into a proton (producing other particles) as described above; the reverse can happen inside other nuclides, where a proton turns into a neutron (producing other particles) through β^+ decay, or electron capture. And inside still other nuclides, both protons and neutrons are stable and do not change form.

Antinucleons

Both nucleons have corresponding antiparticles: the antiproton and the antineutron, which have the same mass and opposite charge as the proton and neutron respectively, and they interact in the same way. (This is generally believed to be *exactly* true, due to CPT symmetry. If there is a difference, it is too small to measure in all experiments to date.) In particular, antinucleons can bind into an "antinucleus". So far, scientists have created antideuterium and antihelium-3 nuclei.

Meson

Meson, any member of a family of subatomic particles composed of a quark and an antiquark. Mesons are sensitive to the strong force, the fundamental interaction that binds the components of the nucleus by governing the behaviour of their constituent quarks. Predicted theoretically in 1935 by the Japanese physicist Yukawa Hideki, the existence of mesons was confirmed in 1947 by a team led by the English physicist Cecil Frank Powell with the discovery of the pi-meson (pion) in cosmic-rayparticle interactions. More than 200 mesons have been produced and characterized in the intervening years, most in high-energy particle-accelerator experiments. All mesons are unstable, with lifetimes ranging from 10^{-8} second to less than 10^{-22} second. They also vary widely in mass, from 140 megaelectron volts (MeV; 10^6 eV) to nearly 10 gigaelectron volts (GeV; 10^9 eV). Mesons serve as a useful tool for studying the properties and interactions of quarks.

Despite their instability, many mesons last long enough (a few billionths of a second) to be observed with particle detectors, making it possible for researchers to reconstruct the motions of quarks. Any model attempting to explain quarks must correctly elucidate the behaviour of mesons. One of the early successes of the Eightfold Way—a forerunner of modern quark models devised by the physicists Murray Gell-Mann and Yuval Ne'eman—was the prediction and subsequent discovery of the eta-meson (1962). Some years later the decay rate of the pi-meson into two photons was used to support the hypothesis that quarks can take on one of three "colours." Studies of the competing decay modes of K-mesons, which occur via the weak force, have led to a better understanding of parity (the property of an elementary particle or physical system that indicates whether its mirror image occurs in nature) and its nonconservation in the weak interaction. CP violation(the violation of the combined conservation lawsassociated with charge and parity) was discovered first in the K-meson system and is under investigation in B-mesons (which contain bottom quarks).

Mesons also provide a means of identifying new quarks. The J/psi particle, discovered independently by teams led by the American physicists Samuel C.C. Ting and Burton Richter in 1974, proved to be a meson made up of a charm quark and its antiquark. (Up to this time, three quark types had been postulated—up, down, and strange.) It was the first manifestation of charm, a new quantum number the existence of which implies that quarks are related in pairs. The subsequent discovery of another heavy meson, called upsilon, revealed the existence of the bottom quarkand its accompanying antiquark and gave rise to speculation about the existence of a companion particle, the top quark. This sixth quark type, or "flavour," was discovered in 1995. Conclusive proof of its existence culminated the search for one of the last missing pieces in the Standard Model of particle physics, which describes the fundamental particles and their interactions.

Spin, Orbital Angular Momentum and Total Angular Momentum

Spin (quantum number S) is a vector quantity that represents the "intrinsic" angular momentum of a particle. It comes in increments of $\frac{1}{2}\hbar$. The $\hbar$ is often dropped because it is the "fundamental" unit of spin, and it is implied that "spin 1" means "spin 1 $\hbar$". (In some systems of natural units, $\hbar$ is chosen to be 1, and therefore does not appear in equations.)

Quarks are fermions—specifically in this case, particles having spin $\frac{1}{2}$ ($S = \frac{1}{2}$). Because spin projections vary in increments of 1 (that is 1 $\hbar$), a single quark has a spin vector of length $\frac{1}{2}$, and has two spin projections ($S_z = +\frac{1}{2}$ and $S_z = -\frac{1}{2}$). Two quarks can have their spins aligned, in which case the two spin vectors add to make a vector of length $S = 1$ and three spin projections ($S_z = +1$, $S_z = 0$, and $S_z = -1$), called the spin-1triplet. If two quarks have unaligned spins, the spin vectors add up to make a vector of length S = 0 and only one spin projection ($S_z = 0$), called the spin-0 singlet. Because mesons are made of one quark and one antiquark, they can be found in triplet and singlet spin states.

There is another quantity of quantized angular momentum, called the orbital angular momentum (quantum number L), that comes in increments of 1 $\hbar$, which represent the angular momentum due to quarks orbiting around each other. The total angular momentum (quantum number J) of a particle is therefore the combination of intrinsic angular momentum (spin) and orbital angular momentum. It can take any value from $J = |L - S|$ to $J = |L + S|$, in increments of 1.

Meson angular momentum quantum numbers for L = 0, 1, 2, 3				
S	L	J	P	J^P
0	0	0	−	0^-
	1	1	+	1^+
	2	2	−	2^-
	3	3	+	3^+
1	0	1	−	1^-
	1	2, 1, 0	+	$2^+, 1^+, 0^+$
	2	3, 2, 1	−	$3^-, 2^-, 1^-$
	3	4, 3, 2	+	$4^+, 3^+, 2^+$

Particle physicists are most interested in mesons with no orbital angular momentum ($L = 0$), therefore the two groups of mesons most studied are the $S = 1$; $L = 0$ and $S = 0$; $L = 0$, which corresponds to $J = 1$ and $J = 0$, although they are not the only ones. It is also possible to obtain $J = 1$ particles from $S = 0$ and $L = 1$. How to distinguish between the $S = 1$, $L = 0$ and $S = 0$, $L = 1$ mesons is an active area of research in meson spectroscopy.

Parity

If the universe were reflected in a mirror, most of the laws of physics would be identical—things would behave the same way regardless of what we call "left" and what we call "right". This concept of mirror reflection is called parity (P). Gravity, the electromagnetic force, and the strong interaction all behave in the same way regardless of whether or not the universe is reflected in a mirror, and thus are said to conserve parity (P-symmetry). However, the weak interaction does distinguish "left" from "right", a phenomenon called parity violation (P-violation).

Based on this, one might think that, if the wavefunction for each particle (more precisely, the quantum field for each particle type) were simultaneously mirror-reversed, then the new set of wavefunctions would perfectly satisfy the laws of physics (apart from the weak interaction). It turns out that this is not quite true: In order for the equations to be satisfied, the wavefunctions of certain types of particles have to be multiplied by −1, in addition to being mirror-reversed. Such particle types are said to have *negative* or *odd* parity ($P = -1$, or alternatively $P = -$), whereas the other particles are said to have *positive* or *even* parity ($P = +1$, or alternatively $P = +$).

For mesons, the parity is related to the orbital angular momentum by the relation:

$$P = (-1)^{L+1}$$

where the L is a result of the parity of the corresponding spherical harmonic of the wavefunction. The "+ 1" comes from the fact that, according to the Dirac equation, a quark and an antiquark have opposite intrinsic parities. Therefore, the intrinsic parity of a meson is the product of the intrinsic parities of the quark (+1) and antiquark (−1). As these are different, their product is −1, and so it contributes the "+ 1" that appears in the exponent.

As a consequence, all mesons with no orbital angular momentum ($L = 0$) have odd parity ($P = -1$).

C-parity

C-parity is only defined for mesons that are their own antiparticle (i.e. neutral mesons). It represents whether or not the wavefunction of the meson remains the same under the interchange of their quark with their antiquark. If

$$|q\bar{q}\rangle = |\bar{q}q\rangle$$

then, the meson is "C even" (C = +1). On the other hand, if

$$|q\bar{q}\rangle = -|\bar{q}q\rangle$$

then the meson is "C odd" (C = −1).

C-parity rarely is studied on its own, but more commonly in combination with P-parity into CP-parity. CP-parity was thought to be conserved, but was later found to be violated in weak interactions.

G-parity

G parity is a generalization of the C-parity. Instead of simply comparing the wavefunction after exchanging quarks and antiquarks, it compares the wavefunction after exchanging the meson for the corresponding antimeson, regardless of quark content.

If

$$|q_1\bar{q}_2\rangle = |\bar{q}_1 q_2\rangle$$

then, the meson is "G even" (G = +1). On the other hand, if

$$|q_1\bar{q}_2\rangle = -|\bar{q}_1 q_2\rangle$$

then the meson is "G odd" (G = −1).

Isospin and Charge

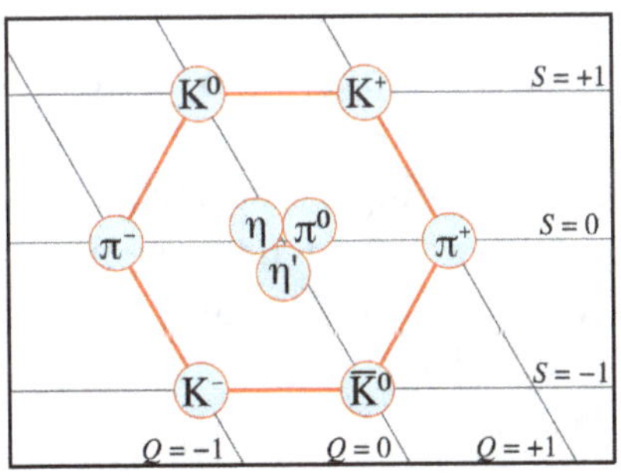

Combinations of one u, d or s quarks and one u, d, or s antiquark in $J^P = 0^-$ configuration form a nonet.

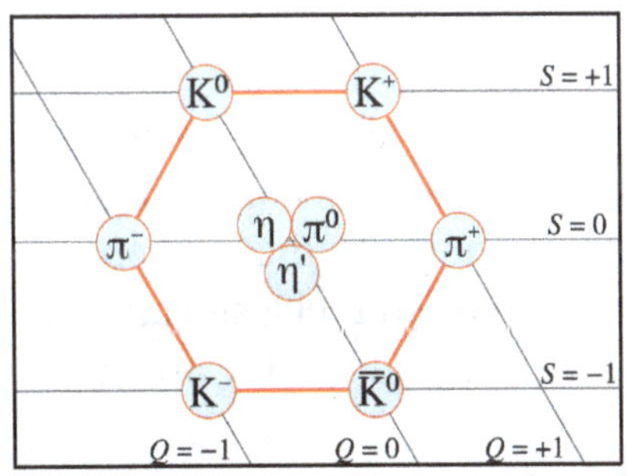

Combinations of one u, d or s quarks and one u, d, or s antiquark in $J^P = 1^-$ configuration also form a nonet.

The concept of isospin was first proposed by Werner Heisenberg in 1932 to explain the similarities between protons and neutrons under the strong interaction. Although they had different electric charges, their masses were so similar that physicists believed that they were actually the same particle. The different electric charges were explained as being the result of some unknown excitation similar to spin. This unknown excitation was later dubbed *isospin* by Eugene Wigner in 1937. When the first mesons were discovered, they too were seen through the eyes of isospin and so the three pions were believed to be the same particle, but in different isospin states.

This belief lasted until Murray Gell-Mann proposed the quark model in 1964 (containing originally only the u, d, and s quarks). The success of the isospin model is now understood to be the result of the similar masses of the u and d quarks. Because the u and d quarks have similar masses, particles made of the same number of them also have similar masses. The exact specific u and d quark composition determines the charge, because u quarks carry charge $+2/3$ whereas d quarks carry charge $-1/3$. For example, the three pions all have different charges (π^+ ($u\bar{d}$), π^0 (a quantum superposition of $u\bar{u}$ and

$d\bar{d}$ states), π^- ($d\bar{u}$)), but have similar masses (c. 140 MeV/c^2) as they are each made of a same number of total of up and down quarks and antiquarks. Under the isospin model, they were considered to be a single particle in different charged states.

The mathematics of isospin was modeled after that of spin. Isospin projections varied in increments of 1 just like those of spin, and to each projection was associated a "charged state". Because the "pion particle" had three "charged states", it was said to be of isospin $I = 1$. Its "charged states" π^+, π^0, and π^-, corresponded to the isospin projections $I_3 = +1$, $I_3 = 0$, and $I_3 = -1$ respectively. Another example is the "rho particle", also with three charged states. Its "charged states" ρ^+, ρ^0, and ρ^-, corresponded to the isospin projections $I_3 = +1$, $I_3 = 0$, and $I_3 = -1$ respectively. It was later noted that the isospin projections were related to the up and down quark content of particles by the relation

$$I_3 = \frac{1}{2}[(n_u - n_{\bar{u}}) - (n_d - n_{\bar{d}})],$$

where the n's are the number of up and down quarks and antiquarks.

In the "isospin picture", the three pions and three rhos were thought to be the different states of two particles. However, in the quark model, the rhos are excited states of pions. Isospin, although conveying an inaccurate picture of things, is still used to classify hadrons, leading to unnatural and often confusing nomenclature. Because mesons are hadrons, the isospin classification is also used, with $I_3 = +\frac{1}{2}$ for up quarks and down antiquarks, and $I_3 = -\frac{1}{2}$ for up antiquarks and down quarks.

Flavour Quantum Numbers

The strangeness quantum number S was noticed to go up and down along with particle mass. The higher the mass, the lower the strangeness (the more s quarks). Particles could be described with isospin projections (related to charge) and strangeness (mass). As other quarks were discovered, new quantum numbers were made to have similar description of udc and udb nonets. Because only the u and d mass are similar, this description of particle mass and charge in terms of isospin and flavour quantum numbers only works well for the nonets made of one u, one d and one other quark and breaks down for the other nonets (for example ucb nonet). If the quarks all had the same mass, their behaviour would be called *symmetric*, because they would all behave in exactly the same way with respect to the strong interaction. However, as quarks do not have the same mass, they do not interact in the same way (exactly like an electron placed in an electric field will accelerate more than a proton placed in the same field because of its lighter mass), and the symmetry is said to be broken.

It was noted that charge (Q) was related to the isospin projection (I_3), the baryon number (B) and flavour quantum numbers (S, C, B', T) by the Gell-Mann–Nishijima formula:

$$Q = I_3 + \frac{1}{2}(B + S + C + B' + T),$$

where S, C, B', and T represent the strangeness, charm, bottomness and topnessflavour quantum numbers respectively. They are related to the number of strange, charm, bottom, and top quarks and antiquark according to the relations:

$$S = -(n_s - n_{\bar{s}})$$

$$C = +(n_c - n_{\bar{c}})$$

$$B' = -(n_b - n_{\bar{b}})$$

$$T = +(n_t - n_{\bar{t}}),$$

meaning that the Gell-Mann–Nishijima formula is equivalent to the expression of charge in terms of quark content:

$$Q = \frac{2}{3}[(n_u - n_{\bar{u}}) + (n_c - n_{\bar{c}}) + (n_t - n_{\bar{t}})] - \frac{1}{3}[(n_d - n_{\bar{d}}) + (n_s - n_{\bar{s}}) + (n_b - n_{\bar{b}})].$$

Classification

Mesons are classified into groups according to their isospin (*I*), total angular momentum(*J*), parity (*P*), G-parity (*G*) or C-parity (*C*) when applicable, and quark (q) content. The rules for classification are defined by the Particle Data Group, and are rather convoluted.

Types of Meson

Mesons are classified into types according to their spin configurations. Some specific configurations are given special names based on the mathematical properties of their spin configuration.

Types of mesons					
Type	***S***	***L***	***P***	***J***	***J^P***
Pseudoscalar meson	0	0	–	0	0^-
Pseudovector meson	0, 1	1	+	1	1^+
Vector meson	1	0, 2	–	1	1^-
Scalar meson	1	1	+	0	0^+
Tensor meson	1	1, 3	+	2	2^+

Nomenclature

Flavourless Mesons

Flavourless mesons are mesons made of pair of quark and antiquarks of the same flavour (all their flavour quantum numbers are zero: $S = 0, C = 0, B' = 0, T = 0$). The rules for flavourless mesons are:

Nomenclature of flavourless mesons

$q\bar{q}$ content	$J^{PC\dagger}$ → I ↓	$0^{-+}, 2^{-+}, 4^{-+}, \ldots$	$1^{+-}, 3^{+-}, 5^{+-}, \ldots$	$1^{--}, 2^{--}, 3^{--}, \ldots$	$0^{++}, 1^{++}, 2^{++}, \ldots$
$u\bar{d}$ $\frac{u\bar{u}-d\bar{d}}{\sqrt{2}}$ $d\bar{u}$	1	π^+ π^0 π^-	b^+ b^0 b^-	ρ^+ ρ^0 ρ^-	a^+ a^0 a^-
Mix of $u\bar{u}$, $d\bar{d}$, $s\bar{s}$	0	η η'	h h′	ω ϕ	f f′
$c\bar{c}$	0	η_c	h_c	$\psi^{\dagger\dagger}$	χ_c
$b\bar{b}$	0	η_b	h_b	Υ	χ_b
$t\bar{t}$	0	η_t	h_t	θ	χ_t

† ^ The C parity is only relevant to neutral mesons.

†† ^ For $J^{PC}=1^{--}$, the ψ is called the J/ψ

In addition:

- When the spectroscopic state of the meson is known, it is added in parentheses.
- When the spectroscopic state is unknown, mass (in MeV/c^2) is added in parentheses.
- When the meson is in its ground state, nothing is added in parentheses.

Flavoured Mesons

Flavoured mesons are mesons made of pair of quark and antiquarks of different flavours. The rules are simpler in this case: the main symbol depends on the heavier quark, the superscript depends on the charge, and the subscript (if any) depends on the lighter quark. In table form, they are:

Nomenclature of flavoured mesons

antiquark → quark ↓	up	down	charm	strange	top	bottom
up	—	[21]	$\bar{D}^0$	K^+	$\bar{T}^0$	B^+
down	[21]	—	D^-	K^0	T^-	B^0
charm	D^0	D^+	—	D_s^+	$\bar{T}_c^0$	B_c^+
strange	K^-	$\bar{K}^0$	D_s^-	—	T_s^-	B_s^0
top	T^0	T^+	T_c^0	T_s^+	—	T_b^+
bottom	B^-	$\bar{B}^0$	B_c^-	$\bar{B}_s^0$	T_b^-	—

In addition:

- If J^P is in the "normal series" (i.e., $J^P = 0^+, 1^-, 2^+, 3^-, \ldots$), a superscript ∗ is added.
- If the meson is not pseudoscalar ($J^P = 0^-$) or vector ($J^P = 1^-$), J is added as a subscript.
- When the spectroscopic state of the meson is known, it is added in parentheses.
- When the spectroscopic state is unknown, mass (in MeV/c^2) is added in parentheses.
- When the meson is in its ground state, nothing is added in parentheses.

Exotic Mesons

There is experimental evidence for particles that are hadrons (i.e., are composed of quarks) and are color-neutral with zero baryon number, and thus by conventional definition are mesons. Yet, these particles do not consist of a single quark/antiquark pair, as all the other conventional mesons discussed above do. A tentative category for these particles is exotic mesons.

There are at least five exotic meson resonances that have been experimentally confirmed to exist by two or more independent experiments. The most statistically significant of these is the Z(4430), discovered by the Belle experiment in 2007 and confirmed by LHCb in 2014. It is a candidate for being a tetraquark: a particle composed of two quarks and two antiquarks.

Mesons

Particle	Symbol	Anti-particle	Makeup	Rest mass MeV/c²	S	C	B	Lifetime	Decay Modes
Pion	π^+	π^-	ud	139.6	0	0	0	2.60 x10^{-8}	$\mu^+\nu_\mu$
Pion	π^0	Self	$\frac{u\bar{u}-d\bar{d}}{\sqrt{2}}$	135.0	0	0	0	0.83 x10^{-16}	2γ
Kaon	K^+	K^-	us	493.7	+1	0	0	1.24 x10^{-8}	$\mu^+\nu_\mu$, $\pi^+\pi^0$
Kaon	K^0_s	K^0_s	1*	497.7	+1	0	0	0.89 x10^{-10}	$\pi^+\pi^-$,$2\pi^0$
Kaon	K^0_L	K^0_L	1*	497.7	+1	0	0	5.2 x10^{-8}	$\pi^+e^-\nu_e$
Eta	η^0	Self	2*	548.8	0	0	0	<10^{-18}	2γ, 3μ
Eta prime	$\eta^{0'}$	Self	2*	958	0	0	0	...	$\pi^+\pi^-\eta$
Rho	ρ^+	ρ^-	ud	770	0	0	0	0.4 x10^{-23}	$\pi^+\pi^0$

Rho	ρ^o	Self	uu, dd	770	0	0	0	0.4 $\times10^{-23}$	$\pi^+\pi^-$
Omega	ω^o	Self	uu, dd	782	0	0	0	0.8 $\times10^{-22}$	$\pi^+\pi^-\pi^o$
Phi	ϕ	Self	ss	1020	0	0	0	20 $\times10^{-23}$	K^+K^-,K^oK^o
D	D^+	D^-	cd	1869.4	0	+1	0	10.6 $\times10^{-13}$	K + _, e + _
D	D^o	D^o	cu	1864.6	0	+1	0	4.2 $\times10^{-13}$	[K,μ,e] + _
D	D^+_s	D^-_s	cs	1969	+1	+1	0	4.7 $\times10^{-13}$	K + _
J/Psi	J/ψ	Self	cc	3096.9	0	0	0	0.8 $\times10^{-20}$	e^+e^-, $\mu^+\mu^-$...
B	B^-	B^+	bu	5279	0	0	-1	1.5 $\times10^{-12}$	D^o + _
B	B^o	B^o	db	5279	0	0	-1	1.5 $\times10^{-12}$	D^o + _
B_s	B_s^o	B_s^o	sb	5370	-1	0	-1	...	B^-_s + _
Upsilon	ϒ	Self	bb	9460.4	0	0	0	1.3 $\times10^{-20}$	e^+e^-, $\mu^+\mu^-$..

Note that an underline indicates an antiparticle.

Color Charge

Color charge is the 3-valued hidden quantum number carried by quarks, antiquarks and gluons. Color charge has a 3 valuedness that we associate with the group $SU(3)_{color}$. Color charge is hidden in the sense that only singlets of $SU(3)_{color}$ that are *neutral* occur in nature (at least macroscopically and at low temperatures). The strongly interacting color-neutral particles composed of quarks, antiquarks and gluons that occur in nature are called *hadrons*. (The word *color* in this context is purely colloquial and has no relation to the color that we see with our eyes in everyday life.)

Color charge has two aspects: (a) as a quantum number that labels states of quarks, antiquarks and gluons: hadrons are in the singlet of $SU(3)_{color}$ as a global symmetry group and (b) as the source of the strong color force acting between quarks associated with $SU(3)_{color}$ as a local gauge group. Each of these is analogous to aspects of electric charge: (a) as a quantum number that counts the amount of electric charge in a state: neutral atoms have zero electric charge under U(1) as a global symmetry group, (b) as the source of electromagnetic forces associated with U(1) as a local gauge group acting between electrically charged particles.

O.W. Greenberg introduced the aspect of color charge as a quantum number in 1964 (Greenberg 1964). Y. Nambu, (Nambu 1966) and M.-Y. Han and Y. Nambu (Han and Nambu 1965) introduced the aspect of color charge as the source of the force between quarks in 1965 associated with the local gauge group $SU(3)_{color}$.

Red, Green and Blue

In quantum chromodynamics, a quark's color can take one of three values or charges: red, green, and blue. An antiquark can take one of three anticolors: called antired, antigreen, and antiblue (represented as cyan, magenta and yellow, respectively). Gluons are mixtures of two colors, such as red and antigreen, which constitutes their color charge. QCD considers eight gluons of the possible nine color–anticolor combinations to be unique.

The following illustrates the coupling constants for color-charged particles:

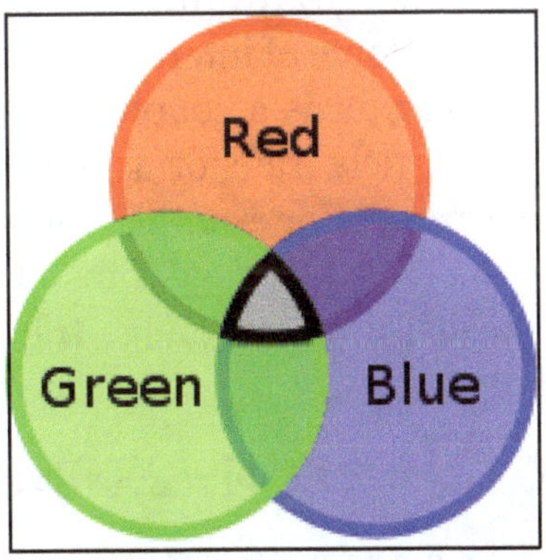

The quark colors (red, green, blue) combine to be colorless.

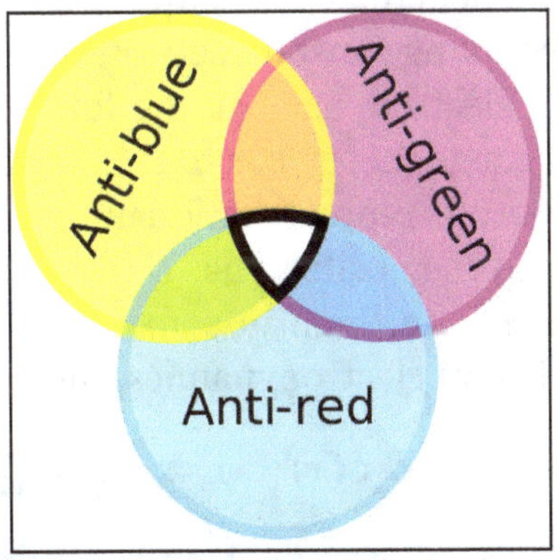

The quark anticolors (antired, antigreen, antiblue) also combine to be colorless.

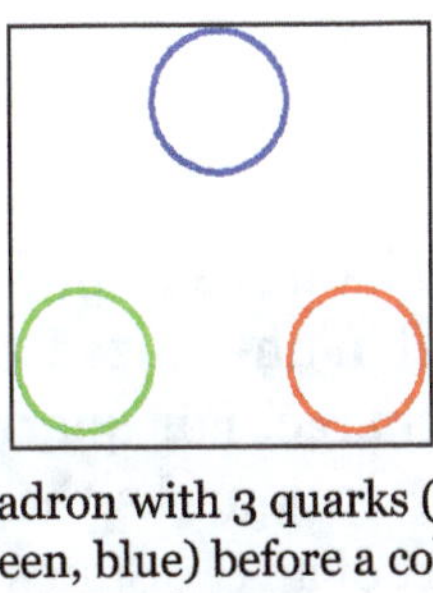

A hadron with 3 quarks (red, green, blue) before a color change.

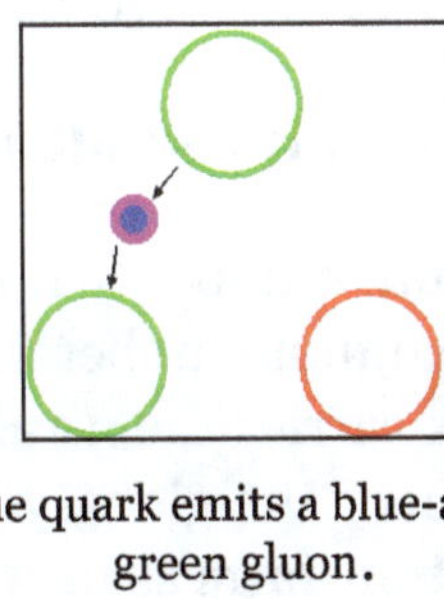

Blue quark emits a blue-antigreen gluon.

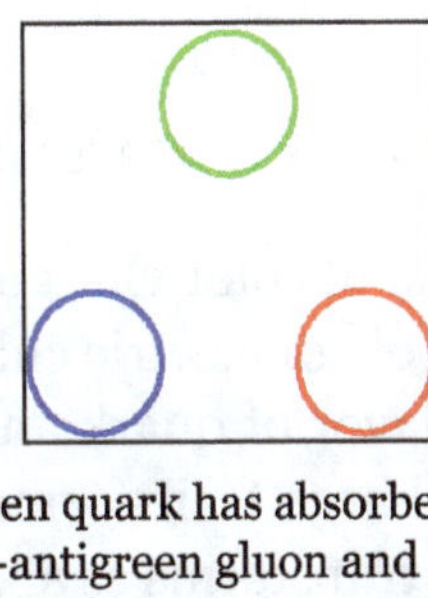

Green quark has absorbed the blue-antigreen gluon and is now blue; color remains conserved.

Field Lines from Color Charges

Analogous to an electric field and electric charges, the strong force acting between color charges can be depicted using field lines. However, the color field lines do not arc outwards from one charge to another as much, because they are pulled together tightly by gluons (within 1 fm). This effect confines quarks within hadrons.

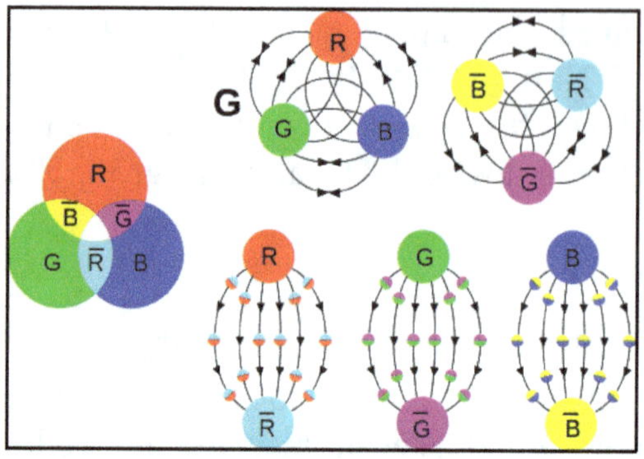

Fields due to color charges, as in quarks (G is the gluon field strength tensor). These are "colorless" combinations. Top: Color charge has "ternary neutral states" as well as binary neutrality (analogous to electric charge). Bottom: Quark/antiquark combinations.

Coupling Constant and Charge

In a quantum field theory, a coupling constant and a charge are different but related notions. The coupling constant sets the magnitude of the force of interaction; for example, in quantum electrodynamics, the fine-structure constant is a coupling constant. The charge in a gauge theory has to do with the way a particle transforms under the gauge symmetry; i.e., its representation under the gauge group. For example, the electron has charge −1 and the positron has charge +1, implying that the gauge transformation has opposite effects on them in some sense. Specifically, if a local gauge transformation $\phi(x)$ is applied in electrodynamics, then one finds:

$$A_\mu \to A_\mu + \partial_\mu \phi(x)\,,\ \ \psi \to \exp[+iQ\phi(x)]\psi \ \text{ and } \ \overline{\psi} \to \exp[-iQ\phi(x)]\overline{\psi}$$

where A_μ is the photon field, and ψ is the electron field with $Q = -1$ (a bar over ψ denotes its antiparticle – the positron). Since QCD is a non-abelian theory, the representations, and hence the color charges, are more complicated.

Color as a Quantum Number that Labels States

The role of color charge as a quantum number that is neutral in hadrons is analogous to the role of electric charge as a quantum number that is neutral in un-ionized atoms. Each flavor of quark and antiquark carries the 3-valued color charge. For quarks the color charge transforms as a 3_α under the global $SU(3)_{color}$; for antiquarks the color charge transforms as a $3^{*\beta}$. Gluons transform as the traceless part of an 8^β_α. States composed of products of quarks, antiquarks and gluons can be reduced into irreducible representations of $SU(3)_{color}$.

The only states that occur macroscopically (and at low temperature) are the singlets of $SU(3)_{color}$. These are hadrons, the color neutral states. For example, mesons have a term $q_\alpha \overline{q}^\alpha$ as the leading constituent, as well as higher components with additional gluons and quark-antiquark pairs such as $q_\alpha G^\alpha_\beta \overline{q}^\beta, q_\alpha \overline{q}^\beta q_\beta \overline{q}^\alpha$ and $q_\alpha \overline{q}^\beta G^\gamma_\beta q_\gamma \overline{q}^\delta G^\alpha_\delta$ constructed so that the state is a color singlet. Baryons have a leading term $\epsilon^{\alpha\beta\gamma}q_\alpha q_\beta q_\gamma$ as well as terms with additional gluons, quarks and antiquarks such as $\epsilon^{\beta\delta\sigma}q_\alpha G^\alpha_\beta q_\gamma G^\gamma_\delta q_\sigma$.

Color as the Source of the Strong Color Force

Color charge plays a second role in connection with the strong force between color-charge carrying particles, just as electric charge plays a second role in connection with the electromagnetic force between electrically charged particles. The electric force is mediated by photons, which are the quanta of the electromagnetic field associated with the local gauge group U(1). Photons do not carry electric charge. In contrast, the strong or color force is associated with the nonabelian local gauge group $SU(3)_{color}$. Because the color charges do not commute with each other, i.e. are nonabelian, the gluons carry color charges, roughly equivalent to the combined charges of a quark and an antiquark.

Quarks carry the the fundamental 3 representation of $SU(3)_{color}$. Antiquarks carry the complex conjugate 3* representation of the group. Gluons, the mediators of the color force, carry the 8-dimensional adjoint representation of the color group. This adjoint representation is analogous to the traceless product of the $SU(3)_{color}$ representations of a quark and an antiquark. The strong force is mediated by the gluons which are quanta of the color or quantum chromodynamic field associated with the local gauge group $SU(3)_{color}$ in analogy to photons as mediators of the electromagnetic force associated with the U(1) gauge theory. However, the gluons interact directly with each other as well as with quarks and antiquarks, in contrast to photons, which interact directly only with electrically charged particles.

Color Force as a Paradigm Shift from Meson Forces as the Source of Strong Interactions

The discovery of quark color in 1964 and the gauge theory, $SU(3)_{color}$, in 1965 changed our understanding of the strong interaction in a qualitative way. Before the discovery of the color carried by quarks, the strong interaction was thought to be mediated by the exchange of mesons, such as the $\pi^{+,o,-}$, and the ω^o, $\rho^{+,o,-}$, ϕ o mesons. With discoveries of quarks and color we now understand the strong interaction to be connected with the $SU(3)_{color}$ gauge theory. We do not believe, however, that single exchange of gluons, which might be expected from perturbation theory, is the main mechanism. Indeed, meson exchange may also play a role in the strong interaction.

Flavor Independence of Color

Hadrons are singlets (i.e., neutral) under the color group, so their color charges are hidden. The coupling of quarks to gluons is independent of the flavor of the quark. Further, the coupling of quarks to photons is also independent of the electric charge of the quark. This latter property is required for the exact conservation of both color charge and electric charge. As stated above, quarks carry fractional values of the electric charge, e, of the proton; however hadrons, which only occur in color singlets, carry integer values of electric charge.

Consequences of the Nonabelian Color Force

The non-abelian character of the color field has profound consequences for the color force. The color force becomes weak (as the reciprocal of the logarithm of the energy) at high energy or short distance, and it becomes strong at low energy or long distance. The weakness at high energy, called asymptotic freedom, provides a justification for the quasi-free behavior of quarks and gluons in the parton model of hadrons, which is useful in describing high-energy scattering. The strength of the color force at low energy leads to the confinement of quarks and gluons, discussed above. This provides an explanation for the observed absence of particles with fractional values of electric charge, despite the fractional values of electric charge carried by quarks.

Road from Color Charge to Quantum Chromodynamics (QCD)

Crucial steps along this road include the demonstration of renormalizability for non-abelian gauge theories, the discovery that these theories are asymptotically free, and the importance of effects outside the realm of perturbation theory.

Empirical Evidence for Color Charge

Ground State Baryons

The first observational evidence for color charge was the permutation symmetry of the quark quantum numbers in the ground state baryons. The $SU(6)$ model of F. Gürsey and L.A. Radicati (Gürsey and Radicati 1964) discussed below placed the ground state baryons in the 56 representation of $SU(6)$. This representation is symmetric under permutations of the quark quantum numbers. Since the quarks have spin 1/2 they should be fermions according to the spin-statistics theorem and should be antisymmetric under permutations of their quantum numbers. This contradiction is the spin-statistics paradox. The introduction of the three-valued hidden color charge by Greenberg in 1964 resolved this contradiction by making the state of the 3-valued color charge carried by quarks antisymmetric. It justified the symmetry of the baryons under permutations of their visible degrees of freedom, which are the space, spin and flavor degrees of freedom.

The Symmetric Quark Model for Baryons

Greenberg constructed the *symmetric quark model* for baryons using the antisymmetry of the color degree of freedom to justify symmetry of the baryon wave functions in the space, spin and flavor degree of freedom. He gave a table of baryon states for quarks in the s and p states. With M. Resnikoff (Greenberg and Resnikoff 1967) he gave a detailed analysis of baryons in the $(56,1^{+})$ and $(70,1^{-})$ supermultiplets, where the notation is (dimSU(6),L^P). This model is the starting point for the study of baryon spectroscopy. For mesons, since there is only one quark or antiquark in the constituent model, the statistics of the quarks is irrelevant.

Neutral Pion Decay to Photons

Further observational evidence for color charge came from the decay rate for $\pi^0 \to \gamma\gamma$ via the axial anomaly. The 3-valued color charge enters as the square of the number of colors and provides a factor of 9 that brings the theoretical calculation of the decay rate into agreement with experiment.

Electron-positron Annihilation to Hadrons

Another experimental evidence came from the ratio $\sigma\left(e^+ e^- \to hadrons\right) / \sigma\left(e^+ e^- \to \mu^+ \mu^-\right)$. Here the 3-valued color charge provides a factor of 3 that brings theory in agreement with experiment. These two tests of color, as well as the permutation symmetry of the quarks in baryons, test the color quantum numbers of the quarks, but do not (in lowest order) test the nonabelian nature of the color force.

Empirical Tests of the Nonabelian Color Force

Among the tests of the nonabelian nature of the color force are asymptotic freedom, permanent confinement of quarks, jets in high-energy scattering, and many precision tests of the standard model. Asymptotic freedom, the decrease of the color force at short distance or high energy, results from the nonabelian interaction of gluons with quarks. Virtual gluons antishield the color charge of a quark (or antiquark) at short distance and thus reduce the color force at short distances. At long distance the growing potential interaction energy of quarks and antiquarks prevents the isolation of individual quarks or antiquarks. Jets in high-energy scattering reflect the transient production of quarks which hadronize to produce strongly interacting particles. Precision tests of quantum chromodynamics, such as the running of coupling constants, also depend on the nonabelian nature of the color interaction.

Historical Developments that Led to Color

Precursors of Quarks

The idea of constructing known particles out of other, more basic particles, has a long history, going back to the effort of M. Born and N. Nagendra Nath (Born and Nagendra Nath 1936) to construct the photon from neutrinos in the 1930's and the effort of E. Fermi and C.N. Yang (Fermi and Yang 1949) to construct pions from protons and neutrons in the 1040's. S. Sakata (Sakata 1956) proposed a model in 1956 with the proton, neutron and lambda baryon were taken as the basic particles. The Sakata model failed to account for the octet of baryons $(p, n, \Lambda, ^{\Sigma +,0,-}, \Xi^{0,-})$.

Quarks

Current Quarks

M. Gell-Mann (Gell-Mann 1964) suggested a model with three *quarks,* $u^{\frac{2}{3}}, d^{-\frac{1}{3}}, s^{-\frac{1}{3}}$ as the fundamental objects in 1964. Gell-Mann's model has the radical departure that the quark electric charges, shown as superscripts above, are fractions of the proton charge e. In addition the quarks carry baryon numbers 1/3 of the baryon number of the nucleon. Particles with either of these fractional values have never been observed. Gell-Mann gave expressions for the vector electromagnetic current and the vector and axial weak currents in terms of quark fields. He showed that quark currents could give the Cabibbo model of weak interactions and also build up the SU(3)×SU(3) current algebra. Gell-Mann's original model used the three *flavors,* u,d,s that were known in 1964. Three additional flavors, c,t,b have since been discovered.

Constituent Quarks

G. Zweig (Zweig 1964) independently introduced a model with fractional electric and baryon charges with emphasis on the constituent point of view. Zweig called the fundamental objects in his model *aces,* a name that was superseded by quarks. He gave an extensive analysis of the space-time and group theory structure of baryons and mesons on the basis of his model. He gave the distinctions between the predictions of the earlier eightfold way and his model. He also gave a reason for the suppression of the decay $\phi\to\rho\pi$ which is an example of the Okubo, Zweig and Iizuka rule (Okubo 1963, Zweig 1964, Iizuka 1966) that can be described in terms of quark line diagrams. Zweig adopted the point of view that his aces are real physical particles. Both Gell-Mann and Zweig assumed the quarks have spin 1/2 in order that mesons can be constructed from $(q\bar{q})$ and baryons can be constructed from (qqq) ; however spin did not enter their models in a deeper way.

Incorporation of Spin in the Quark Model

In the same year, 1964, that Gell-Mann and Zweig independently suggested quarks and aces, F. Gürsey and L.A. Radicati (Gürsey and Radicati 1964) incorporated spin in the model in a more intrinsic way. Gürsey and Radicati considered the quarks to be a 6 of an $SU(6)$ group that included the $SU(3)_{flavor}$ of the original quark model with the 3 flavors u,d,s and the $SU(2)_{spin}$ of the quark spin. Because baryons are composed of 3 quarks, Gürsey and Radicati considered the reduction of $6\otimes 6\otimes 6\to 56+70+70+20$ into irreducibles of $SU(6)$. Gürsey and Radicati placed the ground state baryons in the 56 representation of $SU(6)$ which includes the known nucleon octet and delta decuplet

$$56\to(8,1/2)+(10,3/2)\,under\,SU(6)_{fS}\to SU(3)_f\times SU(2)_s,$$

where the nucleon octet is $\left(p^+,n^0,\Lambda^0,\Sigma^+,\Sigma^0,\Sigma^-,\Xi^0,\Xi^-\right)$ and the delta decuplet is $(\Delta^{++},\Delta^+,\Delta^0,\Delta^-,Y_1^{*+},Y_1^{*0},Y_1^{*-},\Xi^{*0},\Xi^{*-},\Omega^-)$. The 56 representation has the three quarks in a symmetric state under permutations in contradiction with what one would expect. The spin-statistics theorem requires that spin 1/2 quarks obey the Pauli exclusion principle and occur in the 20 representation which is antisymmetric under permutations if there are no additional degrees of freedom carried by quarks.

The spin-statistics Paradox

Gürsey and Radicati noted the symmetry of the 56 representation for the ground state baryons and suggested that this would imply that the forces between the quarks are repulsive. R.H. Dalitz (Dalitz 1966) supported the idea that an antisymmetric space wavefunction for the ground state baryons would allow the $SU(3)$ and spin state to be symmetric without violating the spin-statistics connection. Many physicists were skeptical of the reality of quarks. If quarks were merely a mathematical device then perhaps their statistics could be ignored. Greenberg (Greenberg 1964) proposed the solution to the spin-statistics paradox that has been confirmed by experiment and was ultimately accepted by the physics community. He suggested that each flavor of quark comes in three varieties, colloquially called *colors*. The quark color degree of freedom, can be taken treated as the fundamental 3 of a new $SU(3)_{color}$ symmetry. Then in a neutral state, i.e. in the singlet of $SU(3)_{color}$, the color degree of freedom is antisymmetric. The quarks as fermions are then antisymmetric under permutations of all their degrees of freedom. The total antisymmetry comes about from the product of the symmetry of the space, spin and flavor degrees of freedom and the antisymmetry of the color degree of freedom. (Greenberg introduced color using parafermi statistics of order 3 and showed that the generalized spin-statistics theorem for parastatistics is obeyed.)

Skepticism about Quarks and Color

The idea of quarks was received with skepticism because particles with fractional electric charges and baryon number had never been observed. The suggestion that, in addition, quarks carry a new hidden three-valued charge was received with even greater skepticism. Because of this quarks and color were not accepted generally by the physics community until the discovery of the J/Ψ and other hadrons that carry charm quarks in 1974.

For the mesons composed of a quark and an antiquark the quark statistics is irrelevant.

The Relation between Parafermi Statistics and Explicit Color

The relation between the order 3 of parafermi statistics for quarks and the order 3 for the explicit color first proposed by Nambu and by Han and Nambu was clarified by Greenberg and D. Zwanziger in 1966 (Greenberg and Zwanziger 1966). They showed that the states that are bosons or fermions in the parastatistics model are in 1-to1 correspondence with the states that are color singlets in the $SU(3)_{color}$ model. Further analysis of this question was given by Drühl, Haag and Roberts (Drühl, Haag and Roberts 1970).

References

- D.M. Manley (2005). "Status of baryon spectroscopy". Journal of Physics: Conference Series. 5: 230–237. Bibcode:2005JPhCS...9..230M. doi:10.1088/1742-6596/9/1/043
- "Belle Discovers a New Type of Meson" (Press release). KEK. 2007. Archived from the original on 2009-01-22. Retrieved 04, February 2020
- K. Nakamura; et al. (Particle Data Group) (2010). "Review of Particles Physics: The CKM Quark-Mixing Matrix" (PDF). Journal of Physics G. 37: 075021. Bibcode:-2010JPhG...37g5021N. doi:10.1088/0954-3899/37/7A/075021
- J. I. Friedman. "The Road to the Nobel Prize". Huế University. Archived from the original on 2008-12-25. 17, July 2020
- W.-M. Yao; et al. (Particle Data Group) (2006). "Review of Particle Physics: Pentaquark Update" (PDF). Journal of Physics G. 33 (1): 1–1232. arXiv:astro-ph/0601168. Bibcode:-2006JPhG...33....1Y. doi:10.1088/0954-3899/33/1/001
- H. Garcilazo; J. Vijande & A. Valcarce (2007). "Faddeev study of heavy-baryon spectroscopy". Journal of Physics G. 34 (5): 961–976. arXiv:hep-ph/0703257. doi:10.1088/0954-3899/34/5/014

5

The Standard Model

The standard model of particle physics is a theory which seeks to classify all the known elementary particles as well as describe the electromagnetic force, weak nuclear force and strong nuclear force. The topics elaborated in this chapter will help in gaining a better perspective about the standard models of particle physics as well as its mathematical formulation.

The Standard Model of particle physics is a theory concerning the electromagnetic, weak, and strong nuclear interactions, as well as classifying all the subatomic particles known. It was developed throughout the latter half of the 20th century, as a collaborative effort of scientists around the world. The current formulation was finalized in the mid-1970s upon experimental confirmation of the existence of quarks. Since then, discoveries of the top quark (1995), the tau neutrino (2000), and the Higgs boson (2012) have given further credence to the Standard Model. Because of its success in explaining a wide variety of experimental results, the Standard Model is sometimes regarded as the "theory of almost everything".

Although the Standard Model is believed to be theoretically self-consistent and has demonstrated huge and continued successes in providing experimental predictions, it does leave some phenomena unexplained and it falls short of being a complete theory of fundamental interactions. It does not incorporate the full theory of gravitation as described by general relativity, or account for the accelerating expansion of the universe (as possibly described by dark energy). The model does not contain any viable dark matter particle that possesses all of the required properties deduced from observational cosmology. It also does not incorporate neutrino oscillations (and their non-zero masses).

The development of the Standard Model was driven by theoretical and experimental particle physicists alike. For theorists, the Standard Model is a paradigm of a quantum field theory, which exhibits a wide range of physics including spontaneous symmetry breaking, anomalies and non-perturbative behavior. It is used as a basis for building more exotic models that incorporate hypothetical particles, extra dimensions, and elaborate symmetries (such as supersymmetry) in an attempt to explain experimental results at variance with the Standard Model, such as the existence of dark matter and neutrino oscillations.

Historical Background

The first step towards the Standard Model was Sheldon Glashow's discovery in 1961 of a way to combine the electromagnetic and weak interactions. In 1967 Steven Weinberg and Abdus Salam incorporated the Higgs mechanism into Glashow's electroweak interaction, giving it its modern form.

The Higgs mechanism is believed to give rise to the masses of all the elementary particles in the Standard Model. This includes the masses of the W and Z bosons, and the masses of the fermions, i.e. the quarks and leptons.

After the neutral weak currents caused by Z boson exchange were discovered at CERN in 1973, the electroweak theory became widely accepted and Glashow, Salam, and Weinberg shared the 1979 Nobel Prize in Physics for discovering it. The $W^{\pm}$ and Z^{0} bosons were discovered experimentally in 1983; and the ratio of their masses was found to be as the Standard Model predicted.

The theory of the strong interaction (i.e. QCD), to which many contributed, acquired its modern form around 1973–74, when experiments confirmed that the hadrons were composed of fractionally charged quarks.

Overview

At present, matter and energy are best understood in terms of the kinematics and interactions of elementary particles. To date, physics has reduced the laws governing the behavior and interaction of all known forms of matter and energy to a small set of fundamental laws and theories. A major goal of physics is to find the "common ground" that would unite all of these theories into one integrated theory of everything, of which all the other known laws would be special cases, and from which the behavior of all matter and energy could be derived (at least in principle).

Particle Content

The Standard Model includes members of several classes of elementary particles (fermions, gauge bosons, and the Higgs boson), which in turn can be distinguished by other characteristics, such as color charge.

Fermions

The Standard Model includes 12 elementary particles of spin $\frac{1}{2}$ known as fermions. According to the spin-statistics theorem, fermions respect the Pauli exclusion principle. Each fermion has a corresponding antiparticle.

The fermions of the Standard Model are classified according to how they interact (or equivalently, by what charges they carry). There are six quarks (up, down, charm, strange, top, bottom), and six leptons (electron, electron neutrino, muon, muon neutrino, tau, tau neutrino). Pairs from each classification are grouped together to form a generation, with corresponding particles exhibiting similar physical behavior.

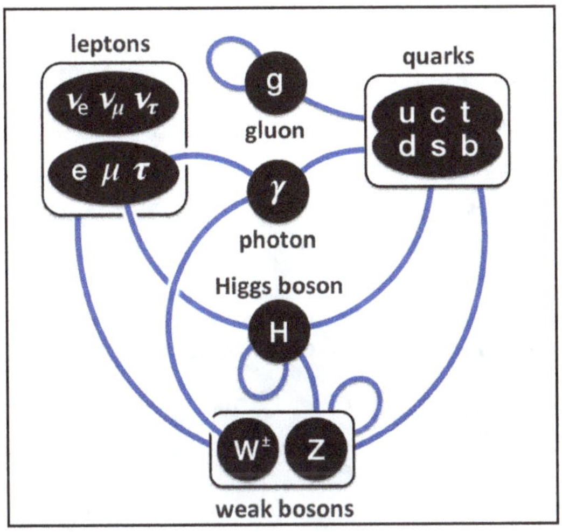

Summary of interactions between particles described by the Standard Model.

The defining property of the quarks is that they carry color charge, and hence, interact via the strong interaction. A phenomenon called color confinement results in quarks being very strongly bound to one another, forming color-neutral composite particles (hadrons) containing either a quark and an antiquark (mesons) or three quarks (baryons). The familiar proton and the neutron are the two baryons having the smallest mass. Quarks also carry electric charge and weak isospin. Hence, they interact with other fermions both electromagnetically and via the weak interaction.

The remaining six fermions do not carry colour charge and are called leptons. The three neutrinos do not carry electric charge either, so their motion is directly influenced only by the weak nuclear force, which makes them notoriously difficult to detect. However, by virtue of carrying an electric charge, the electron, muon, and tau all interact electromagnetically.

Each member of a generation has greater mass than the corresponding particles of lower generations. The first generation charged particles do not decay; hence all ordinary (baryonic) matter is made of such particles. Specifically, all atoms consist of electrons orbiting around atomic nuclei, ultimately constituted of up and down quarks. Second and third generation charged particles, on the other hand, decay with very short half lives, and are observed only in very high-energy environments. Neutrinos of all generations also do not decay, and pervade the universe, but rarely interact with baryonic matter.

Gauge Bosons

In the Standard Model, gauge bosons are defined as force carriers that mediate the strong, weak, and electromagnetic fundamental interactions.

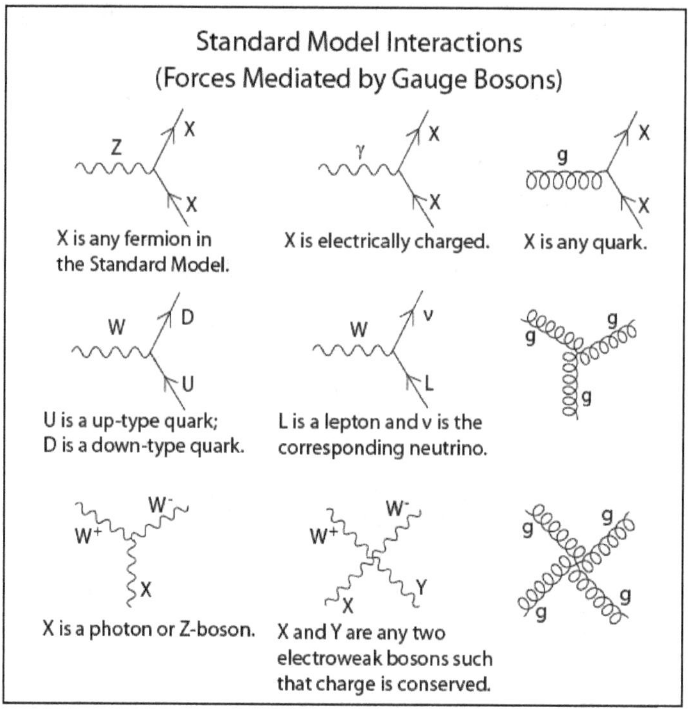

The above interactions form the basis of the standard model. Feynman diagrams in the standard model are built from these vertices. Modifications involving Higgs boson interactions and neutrino oscillations are omitted. The charge of the W bosons is dictated by the fermions they interact with; the conjugate of each listed vertex (i.e. reversing the direction of arrows) is also allowed.

Interactions in physics are the ways that particles influence other particles. At a macroscopic level, electromagnetism allows particles to interact with one another via electric and magnetic fields, and gravitation allows particles with mass to attract one another in accordance with Einstein's theory of general relativity. The Standard Model explains such forces as resulting from matter particles exchanging other particles, generally referred to as *force mediating particles*. When a force-mediating particle is exchanged, at a macroscopic level the effect is equivalent to a force influencing both of them, and the particle is therefore said to have *mediated* (i.e., been the agent of) that force. The Feynman diagram calculations, which are a graphical representation of the perturbation theory approximation, invoke "force mediating particles", and when applied to analyze high-energy scattering experiments are in reasonable agreement with the data. However, perturbation theory (and with it the concept of a "force-mediating particle") fails in other situations. These include low-energy quantum chromodynamics, bound states, and solitons.

The gauge bosons of the Standard Model all have spin (as do matter particles). The value of the spin is 1, making them bosons. As a result, they do not follow the Pauli exclusion principle that constrains fermions: thus bosons (e.g. photons) do not have a theoretical limit on their spatial density (number per volume). The different types of gauge bosons are described below.

- Photons mediate the electromagnetic force between electrically charged particles. The photon is massless and is well-described by the theory of quantum electrodynamics.
- The W+, W−, and Z gauge bosons mediate the weak interactions between particles of different flavors (all quarks and leptons). They are massive, with the Z being more massive than the W±. The weak interactions involving the W± exclusively act on *left-handed* particles and *right-handed* antiparticles. Furthermore, the W± carries an electric charge of +1 and −1 and couples to the electromagnetic interaction. The electrically neutral Z boson interacts with both left-handed particles and antiparticles. These three gauge bosons along with the photons are grouped together, as collectively mediating the electroweak interaction.
- The eight gluons mediate the strong interactions between color charged particles (the quarks). Gluons are massless. The eightfold multiplicity of gluons is labeled by a combination of color and anticolor charge (e.g. red–antigreen). Because the gluons have an effective color charge, they can also interact among themselves. The gluons and their interactions are described by the theory of quantum chromodynamics.

The interactions between all the particles described by the Standard Model are summarized by the diagrams on the right of this section.

Higgs Boson

The Higgs particle is a massive scalar elementary particle theorized by Peter Higgs in 1964 (see 1964 PRL symmetry breaking papers) and is a key building block in the Standard Model. It has no intrinsic spin, and for that reason is classified as a boson (like the gauge bosons, which have integer spin).

The Higgs boson plays a unique role in the Standard Model, by explaining why the other elementary particles, except the photon and gluon, are massive. In particular, the Higgs boson explains why the photon has no mass, while the W and Z bosons are very heavy. Elementary particle masses, and the differences between electromagnetism (mediated by the photon) and the weak force (mediated by the W and Z bosons), are critical to many aspects of the structure of microscopic (and hence macroscopic) matter. In electroweak theory, the Higgs boson generates the masses of the leptons (electron, muon, and tau) and quarks. As the Higgs boson is massive, it must interact with itself.

Because the Higgs boson is a very massive particle and also decays almost immediately when created, only a very high-energy particle accelerator can observe and record it. Experiments to confirm and determine the nature of the Higgs boson using the Large Hadron Collider (LHC) at CERN began in early 2010, and were performed at Fermilab's Tevatron until its closure in late 2011. Mathematical consistency of the Standard

Model requires that any mechanism capable of generating the masses of elementary particles becomes visible at energies above 1.4 TeV; therefore, the LHC (designed to collide two 7 to 8 TeV proton beams) was built to answer the question of whether the Higgs boson actually exists.

On 4 July 2012, the two main experiments at the LHC (ATLAS and CMS) both reported independently that they found a new particle with a mass of about 125 GeV/c^2 (about 133 proton masses, on the order of 10^{-25} kg), which is "consistent with the Higgs boson." Although it has several properties similar to the predicted "simplest" Higgs, they acknowledged that further work would be needed to conclude that it is indeed the Higgs boson, and exactly which version of the Standard Model Higgs is best supported if confirmed.

On 14 March 2013 the Higgs Boson was tentatively confirmed to exist.

Theoretical Aspects

Construction of the Standard Model Lagrangian.

Parameters of the Standard Model			
Symbol	**Description**	**Renormalization scheme (point)**	**Value**
m_e	Electron mass		511 keV
m_μ	Muon mass		105.7 MeV
m_τ	Tau mass		1.78 GeV
m_u	Up quark mass	μ_{MS} = 2 GeV	1.9 MeV
m_d	Down quark mass	μ_{MS} = 2 GeV	4.4 MeV
m_s	Strange quark mass	μ_{MS} = 2 GeV	87 MeV
m_c	Charm quark mass	$\mu_{MS} = m_c$	1.32 GeV
m_b	Bottom quark mass	$\mu_{MS} = m_b$	4.24 GeV
m_t	Top quark mass	On-shell scheme	172.7 GeV
θ_{12}	CKM 12-mixing angle		13.1°
θ_{23}	CKM 23-mixing angle		2.4°
θ_{13}	CKM 13-mixing angle		0.2°
δ	CKM CP-violating Phase		0.995
g_1 or g'	U(1) gauge coupling	$\mu_{MS} = m_Z$	0.357
g_2 or g	SU(2) gauge coupling	$\mu_{MS} = m_Z$	0.652
g_3 or g_s	SU(3) gauge coupling	$\mu_{MS} = m_Z$	1.221
θ_{QCD}	QCD vacuum angle		~0
v	Higgs vacuum expectation value		246 GeV
m_H	Higgs mass		125.36±0.41 GeV (tentative)

Technically, quantum field theory provides the mathematical framework for the Standard Model, in which a Lagrangian controls the dynamics and kinematics of the theory. Each kind of particle is described in terms of a dynamical field that pervades space-time. The construction of the Standard Model proceeds following the modern method of constructing most field theories: by first postulating a set of symmetries of the system, and then by writing down the most general renormalizable Lagrangian from its particle (field) content that observes these symmetries.

The global Poincaré symmetry is postulated for all relativistic quantum field theories. It consists of the familiar translational symmetry, rotational symmetry and the inertial reference frame invariance central to the theory of special relativity. The local SU(3)×SU(2)×U(1) gauge symmetry is an internal symmetry that essentially defines the Standard Model. Roughly, the three factors of the gauge symmetry give rise to the three fundamental interactions. The fields fall into different representations of the various symmetry groups of the Standard Model. Upon writing the most general La-grangian, one finds that the dynamics depend on 19 parameters, whose numerical val-ues are established by experiment. The parameters are summarized in the table above (note: the Higgs mass is at 125 GeV, the Higgs self-coupling strength $\lambda \sim \frac{1}{8}$).

Quantum Chromodynamics Sector

The quantum chromodynamics (QCD) sector defines the interactions between quarks and gluons, with SU(3) symmetry, generated by T^a. Since leptons do not interact with gluons, they are not affected by this sector. The Dirac Lagrangian of the quarks coupled to the gluon fields is given by,

Ga μ is the SU(3) gauge field containing the gluons, γ^μ are the Dirac matrices, D and U are the Dirac spinors associated with up- and down-type quarks, and g_s is the strong coupling constant.

Electroweak Sector

The electroweak sector is a Yang–Mills gauge theory with the simple symmetry group $U(1)\times SU(2)_L$,

$$\mathcal{L}_{EW} = \sum_{\psi} \bar{\psi}\gamma^{\mu}\left(i\partial_{\mu} - g'\frac{1}{2}Y_W B_{\mu} - g\frac{1}{2}\vec{\tau}_L \vec{W}_{\mu}\right)\psi,$$

where B_μ is the U(1) gauge field; Y_W is the weak hypercharge—the generator of the U(1) group; $W\rightarrow_\mu$ is the three-component SU(2) gauge field; $\tau_L\rightarrow$ are the Pauli matrices—infinitesimal generators of the SU(2) group (the subscript L indicates that they only act on left fermions) g' and g are coupling constants.

Higgs Sector

In the Standard Model, the Higgs field is a complex scalar of the group $SU(2)_L$:

$$\varphi = \frac{1}{\sqrt{2}}\begin{pmatrix}\varphi^{+} \\ \varphi^{0}\end{pmatrix},$$

where the indices + and o indicate the electric charge (Q) of the components. The weak isospin (Y_{W}) of both components is 1.

Before symmetry breaking, the Higgs Lagrangian is:

$$\mathcal{L}_{\mathrm{H}} = \varphi^{\dagger}\left(\partial^{\mu} - \frac{i}{2}\left(g' Y_{\mathrm{W}} B^{\mu} + g\vec{\tau}\vec{W}^{\mu}\right)\right)\left(\partial_{\mu} + \frac{i}{2}\left(g' Y_{\mathrm{W}} B_{\mu} + g\vec{\tau}\vec{W}_{\mu}\right)\right)\varphi - \frac{\lambda^{2}}{4}\left(\varphi^{\dagger}\varphi - v^{2}\right)^{2},$$

which can also be written as:

$$\mathcal{L}_{\mathrm{H}} = \left|\left(\partial_{\mu} + \frac{i}{2}\left(g' Y_{\mathrm{W}} B_{\mu} + g\vec{\tau}\vec{W}_{\mu}\right)\right)\varphi\right|^{2} - \frac{\lambda^{2}}{4}\left(\varphi^{\dagger}\varphi - v^{2}\right)^{2}.$$

Fundamental Forces

The Standard Model classified all four fundamental forces in nature. In the Standard Model, a force is described as an exchange of bosons between the objects affected, such as a photon for the electromagnetic force and a gluon for the strong interaction. Those particles are called force carriers.

The four fundamental forces of nature

Property/Interaction	Gravitation	Weak	Electromagnetic	Strong	
		(Electroweak)		Fundamental	Residual
Acts on:	Mass - Energy	Flavor	Electric charge	Color charge	Atomic nuclei
Particles experiencing:	All	Quarks, leptons	Electrically charged	Quarks, Gluons	Hadrons
Particles mediating:	Graviton (theoretical)	W^{+} W^{-} Z^{0}	γ	Gluons	Mesons
Strength in the scale of quarks:	10^{-41}	10^{-4}	1	60	Not applicable to quarks
Strength in the scale of protons/neutrons:	10^{-36}	10^{-7}	1	Not applicable to hadrons	20

Tests and Predictions

The Standard Model (SM) predicted the existence of the W and Z bosons, gluon, and the top and charm quarks before these particles were observed. Their predicted properties were experimentally confirmed with good precision.

The SM also makes several predictions about the decay of Z^0 bosons, which have been experimentally confirmed by the Large Electron-Positron Collider at CERN.

In May 2012 BaBar Collaboration reported that their recently analyzed data may suggest possible flaws in the Standard Model of particle physics. These data show that a particular type of particle decay called "B to D-star-tau-nu" happens more often than the Standard Model says it should. In this type of decay, a particle called the B-bar meson decays into a D meson, an antineutrino and a tau-lepton. While the level of certainty of the excess (3.4 sigma) is not enough to claim a break from the Standard Model, the results are a potential sign of something amiss and are likely to impact existing theories, including those attempting to deduce the properties of Higgs bosons.

On December 13, 2012, physicists reported the constancy, over space and time, of a basic physical constant of nature that supports the *standard model of physics*. The scientists, studying methanol molecules in a distant galaxy, found the change ($\Delta\mu/\mu$) in the proton-to-electron mass ratio μ to be equal to "$(0.0 \pm 1.0) \times 10^{-7}$ at redshift $z = 0.89$" and consistent with "a null result".

Challenges

Unsolved problem in physics:

- What gives rise to the Standard Model of particle physics?
- Why do particle masses and coupling constants have the values that we measure?
- Why are there three generations of particles?
- Why is there more matter than antimatter in the universe?
- Where does Dark Matter fit into the model? Is it even a new particle?

(more unsolved problems in physics)

Self-consistency of the Standard Model (currently formulated as a non-abelian gauge theory quantized through path-integrals) has not been mathematically proven. While regularized versions useful for approximate computations (for example lattice gauge theory) exist, it is not known whether they converge (in the sense of S-matrix elements) in the limit that the regulator is removed. A key question related to the consistency is the Yang–Mills existence and mass gap problem.

Experiments indicate that neutrinos have mass, which the classic Standard Model did not allow. To accommodate this finding, the classic Standard Model can be modified to include neutrino mass.

If one insists on using only Standard Model particles, this can be achieved by adding a non-renormalizable interaction of leptons with the Higgs boson. On a fundamental level, such an interaction emerges in the seesaw mechanism where heavy right-handed neutrinos are added to the theory. This is natural in the left-right symmetric extension of the Standard Model and in certain grand unified theories. As long as new physics appears below or around 10^{14} GeV, the neutrino masses can be of the right order of magnitude.

Theoretical and experimental research has attempted to extend the Standard Model into a Unified field theory or a Theory of everything, a complete theory explaining all physical phenomena including constants. Inadequacies of the Standard Model that motivate such research include:

- The model does not explain gravitation, although physical confirmation of a theoretical particle known as a graviton would account for it to a degree. Though it addresses strong and electroweak interactions, the Standard Model does not consistently explain the canonical theory of gravitation, general relativity, in terms of quantum field theory. The reason for this is, among other things, that quantum field theories of gravity generally break down before reaching the Planck scale. As a consequence, we have no reliable theory for the very early universe.
- Some physicists consider it to be *ad hoc* and inelegant, requiring 19 numerical constants whose values are unrelated and arbitrary. Although the Standard Model, as it now stands, can explain why neutrinos have masses, the specifics of neutrino mass are still unclear. It is believed that explaining neutrino mass will require an additional 7 or 8 constants, which are also arbitrary parameters.
- The Higgs mechanism gives rise to the hierarchy problem if some new physics (coupled to the Higgs) is present at high energy scales. In these cases, in order for the weak scale to be much smaller than the Planck scale, severe fine tuning of the parameters is required; there are, however, other scenarios that include quantum gravity in which such fine tuning can be avoided. There are also issues of Quantum triviality, which suggests that it may not be possible to create a consistent quantum field theory involving elementary scalar particles.
- The model is inconsistent with the emerging "Standard Model of cosmology." More common contentions include the absence of an explanation in the Standard Model for the observed amount of cold dark matter (CDM) and its contributions to dark energy, which are many orders of magnitude too large. It is also difficult to accommodate the observed predominance of matter over antimatter (matter/antimatter asymmetry). The isotropy and homogeneity of the visible universe over large distances seems to require a mechanism like cosmic inflation, which would also constitute an extension of the Standard Model.
- The existence of ultra-high-energy cosmic rays are difficult to explain under the Standard Model.

Currently, no proposed Theory of Everything has been widely accepted or verified.

Mathematical Formulation of the Standard Model

This article describes the mathematics of the Standard Model of particle physics, a gauge quantum field theory containing the internal symmetries of the unitary product group SU(3) × SU(2) × U(1). The theory is commonly viewed as containing the fundamental set of particles – the leptons, quarks, gauge bosons and the Higgs particle.

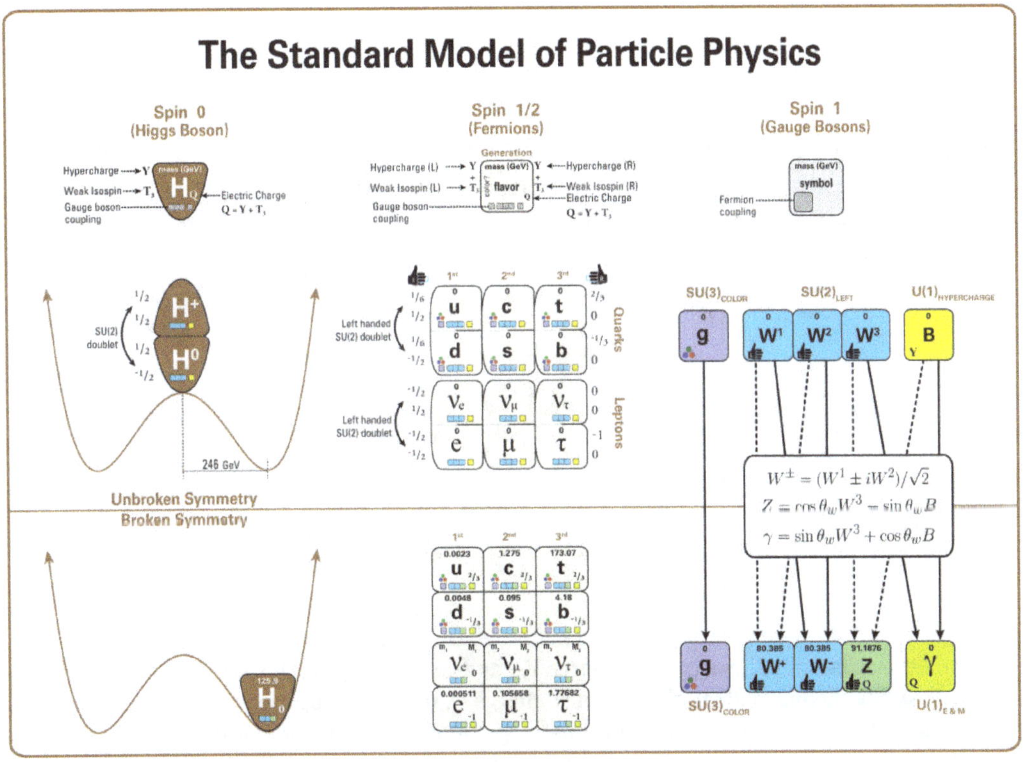

Standard Model of Particle Physics. The diagram shows the elementary particles of the Standard Model (the Higgs boson, the three generations of quarks and leptons, and the gauge bosons), including their names, masses, spins, charges, chiralities, and interactions with the strong, weak and electromagnetic forces. It also depicts the crucial role of the Higgs boson in electroweak symmetry breaking, and shows how the properties of the various particles differ in the (high-energy) symmetric phase (top) and the (low-energy) broken-symmetry phase (bottom).

The Standard Model is renormalizable and mathematically self-consistent, however despite having huge and continued successes in providing experimental predictions it does leave some unexplained phenomena. In particular, although the physics of special relativity is incorporated, general relativity is not, and the Standard Model will fail at energies or distances where the graviton is expected to emerge. Therefore, in a modern field theory context, it is seen as an effective field theory.

This article requires some background in physics and mathematics, but is designed as both an introduction and a reference.

Quantum Field Theory

The standard model is a quantum field theory, meaning its fundamental objects are *quantum fields* which are defined at all points in spacetime. These fields are

- the fermion fields, ψ, which account for "matter particles";
- the electroweak boson fields W_1, W_2, W_3, and B;
- the gluon field, G_a; and
- the Higgs field, φ.

That these are *quantum* rather than *classical* fields has the mathematical consequence that they are operator-valued. In particular, values of the fields generally do not commute. As operators, they act upon the quantum state (ket vector).

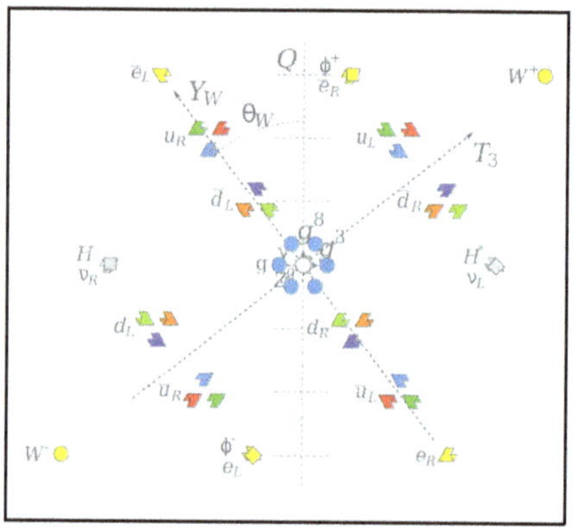

The pattern of weak isospin T_3, weak hypercharge Y_W, and color charge of all known elementary particles, rotated by the weak mixing angle to show electric charge Q, roughly along the vertical. The neutral Higgs field (gray square) breaks the electroweak symmetry and interacts with other particles to give them mass.

The dynamics of the quantum state and the fundamental fields are determined by the Lagrangian density $\mathcal{L}$ (usually for short just called the Lagrangian). This plays a role similar to that of the Schrödinger equation in non-relativistic quantum mechanics, but a Lagrangian is not an equation of motion – rather, it is a polynomial function of the fields and their derivatives, and used with the principle of least action. While it would be possible to derive a system of differential equations governing the fields from the Langrangian, it is more common to use other techniques to compute with quantum field theories.

The standard model is furthermore a gauge theory, which means there are degrees of freedom in the mathematical formalism which do not correspond to changes in the physical state. The gauge group of the standard model is $SU(3) \times SU(2) \times U(1)$, where U(1) acts on B and φ, SU(2) acts on W and φ, and SU(3) acts on G. The fermion field ψ also transforms under these symmetries, although all of them leave some parts of it unchanged.

The Role of the Quantum Fields

In classical mechanics, the state of a system can usually be captured by a small set of variables, and the dynamics of the system is thus determined by the time evolution of these variables. In classical field theory, the *field* is part of the state of the system, so in order to describe it completely one effectively introduces separate variables for every point in spacetime (even though there are many restrictions on how the values of the field "variables" may vary from point to point, for example in the form of field equations involving partial derivatives of the fields).

In quantum mechanics, the classical variables are turned into operators, but these do not capture the state of the system, which is instead encoded into a wavefunction ψ or more abstract ket vector. If ψ is an eigenstate with respect to an operator P, then $P\psi = \lambda\psi$ for the corresponding eigenvalue λ, and hence letting an operator P act on ψ is analogous to multiplying ψ by the value of the classical variable to which P corresponds. By extension, a classical formula where all variables have been replaced by the corresponding operators will behave like an operator which, when it acts upon the state of the system, multiplies it by the analogue of the quantity that the classical formula would compute. The formula as such does however not contain any information about the state of the system; it would evaluate to the same operator regardless of what state the system is in.

Quantum fields relate to quantum mechanics as classical fields do to classical mechanics, i.e., there is a separate operator for every point in spacetime, and these operators do not carry any information about the state of the system; they are merely used to exhibit some aspect of the state, at the point to which they belong. In particular, the quantum fields are *not* wavefunctions, even though the equations which govern their time evolution may be deceptively similar to those of the corresponding wavefunction in a semiclassical formulation. There is no variation in strength of the fields between different points in spacetime; the variation that happens is rather one of phase factors.

Vectors, Scalars, and Spinors

Mathematically it may look as though all of the fields are vector-valued (in addition to being operator-valued), since they all have several components, can be multiplied by matrices, etc., but physicists assign a more specific physical meaning to the word: a vector is something which transforms like a four-vector under Lorentz transformations, and a scalar is something which is invariant under Lorentz transformations. The B, W_j, and G_a fields are all vectors in this sense, so the corresponding particles are said to be vector bosons. The Higgs field φ is a scalar.

The fermion field ψ does transform under Lorentz transformations, but not like a vector should; rotations will only turn it by half the angle a proper vector should. Therefore, these constitute a third kind of quantity, which is known as a spinor.

It is common to make use of abstract index notation for the vector fields, in which case the vector fields all come with a Lorentzian index μ, like so: B^{μ}, W_j^{μ}, and G_a^{μ}. If abstract index notation is used also for spinors then these will carry a spinorial index and the Dirac gamma will carry one Lorentzian and two spinorian indices, but it is more common to regard spinors as column matrices and the Dirac gamma γ_{μ} as a matrix which additionally carries a Lorentzian index. The Feynman slash notation can be used to turn a vector field into a linear operator on spinors, like so: $B = \gamma^{\mu} B_{\mu}$; this may involve raising and lowering indices.

Alternative Presentations of the Fields

As is common in quantum theory, there is more than one way to look at things. At first the basic fields given above may not seem to correspond well with the "fundamental particles" in the chart above, but there are several alternative presentations which, in particular contexts, may be more appropriate than those that are given above.

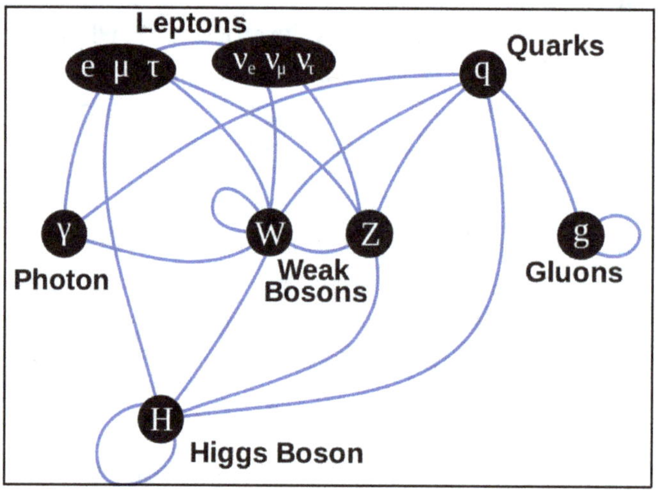

Connections denoting which particles interact with each other.

Fermions

Rather than having one fermion field ψ, it can be split up into separate components for each type of particle. This mirrors the historical evolution of quantum field theory, since the electron component ψ_e (describing the electron and its antiparticle the positron) is then the original ψ field of quantum electrodynamics, which was later accompanied by ψ_{μ} and ψ_{τ} fields for the muon and tauon respectively (and their antiparticles). Electroweak theory added $\psi_{\nu_e}, \psi_{\nu_{\mu}}$, and $\psi_{\nu_{\tau}}$ for the corresponding neutrinos, and the quarks add still further components. In order to be four-spinors like the electron and other lepton components, there must be one quark component for every combination of flavour and colour, bringing the total to 24 (3 for charged leptons, 3 for neutrinos, and 2·3·3 = 18 for quarks). Each of these is a four component bispinor, for a total of 96 complex-valued components for the fermion field.

An important definition is the barred fermion field $\bar{\psi}$ is defined to be $\psi^{\dagger}\gamma^{0}$, where † denotes the Hermitian adjoint and γ^0 is the zeroth gamma matrix. If ψ is thought of as an $n \times 1$ matrix then $\bar{\psi}$ should be thought of as a $1 \times n$ matrix.

A Chiral Theory

An independent decomposition of ψ is that into chirality components:

"Left" chirality: $\psi \quad = -(1 - \gamma \quad)\psi$

"Right" chirality: $\psi^{R} = \frac{1}{2}(1 + \gamma_5)\psi$

where γ_5 is the fifth gamma matrix. This is very important in the Standard Model because *left and right chirality components are treated differently by the gauge interactions.*

In particular, under weak isospin SU(2) transformations the left-handed particles are weak-isospin doublets, whereas the right-handed are singlets – i.e. the weak isospin of ψ_R is zero. Put more simply, the weak interaction could rotate e.g. a left-handed electron into a left-handed neutrino (with emission of a W^-), but could not do so with the same right-handed particles. As an aside, the right-handed neutrino originally did not exist in the standard model – but the discovery of neutrino oscillation implies that neutrinos must have mass, and since chirality can change during the propagation of a massive particle, right-handed neutrinos must exist in reality. This does not however change the (experimentally-proven) chiral nature of the weak interaction.

Furthermore, U(1) acts differently on ψ_{e}^{L} than on ψ_{e}^{R} (because they have different weak hypercharges).

Mass and Interaction Eigenstates

A distinction can thus be made between, for example, the mass and interaction eigenstates of the neutrino. The former is the state which propagates in free space, whereas the latter is the *different* state that participates in interactions. Which is the "fundamental" particle? For the neutrino, it is conventional to define the "flavour" (νe, νμ, or ντ) ( ) by the interaction eigenstate, whereas for the quarks we define the flavour (up, down, etc.) by the mass state. We can switch between these states using the CKM matrix for the quarks, or the PMNS matrix for the neutrinos (the charged leptons on the other hand are eigenstates of both mass and flavour).

As an aside, if a complex phase term exists within either of these matrices, it will give rise to direct CP violation, which could explain the dominance of matter over antimatter in our current universe. This has been proven for the CKM matrix, and is expected for the PMNS matrix.

Positive and Negative Energies

Finally, the quantum fields are sometimes decomposed into "positive" and "negative" energy parts: $\psi = \psi^{+} + \psi^{-}$. This is not so common when a quantum field theory has been set up, but often features prominently in the process of quantizing a field theory.

Bosons

Due to the Higgs mechanism, the electroweak boson fields W_1, W_2, W_3, and B "mix" to create the states which are physically observable. To retain gauge invariance, the underlying fields must be massless, but the observable states can *gain masses* in the process. These states are:

The massive neutral (Z) boson:

$$Z = \cos\theta_W W_3 - \sin\theta_W B$$

The massless neutral boson:

$$A = \sin\theta_W W_3 + \cos\theta_W B$$

The massive charged W bosons:

$$W^{\pm} = \frac{1}{\sqrt{2}}(W_1 \mp i W_2)$$

where θ_W is the Weinberg angle.

The *A* field is the photon, which corresponds classically to the well-known electromagnetic four-potential – i.e. the electric and magnetic fields. The *Z* field actually contributes in every process the photon does, but due to its large mass, the contribution is usually negligible.

Perturbative QFT and the Interaction Picture

Much of the qualitative descriptions of the standard model in terms of "particles" and "forces" comes from the perturbative quantum field theory view of the model. In this, the Langrangian is decomposed as $\mathcal{L} = \mathcal{L}_0 + \mathcal{L}_1$ into separate *free field* and *interaction* Langrangians. The free fields care for particles in isolation, whereas processes involving several particles arise through interactions. The idea is that the state vector should only change when particles interact, meaning a free particle is one whose quantum state is constant. This corresponds to the interaction picture in quantum mechanics.

In the more common Schrödinger picture, even the states of free particles change over time: typically the phase changes at a rate which depends on their energy. In the alternative Heisenberg picture, state vectors are kept constant, at the price of having the operators (in particular the observables) be time-dependent. The interaction picture constitutes an intermediate between the two, where some time dependence is placed in the operators (the quantum fields) and some in the state vector. In QFT, the former is called the free field part of the model, and the latter is called the interaction part. The free field model can be solved exactly, and then the solutions to the full model can be expressed as perturbations of the free field solutions, for example using the Dyson series.

It should be observed that the decomposition into free fields and interactions is in principle arbitrary. For example, renormalization in QED modifies the mass of the free field electron to match that of a physical electron (with an electromagnetic field), and will in doing so add a term to the free field Lagrangian which must be cancelled by a counterterm in the interaction Lagrangian, that then shows up as a two-line vertex in the Feynman diagrams. This is also how the Higgs field is thought to give particles mass: the part of the interaction term which corresponds to the (nonzero) vacuum expectation value of the Higgs field is moved from the interaction to the free field Lagrangian, where it looks just like a mass term having nothing to do with Higgs.

Free Fields

Under the usual free/interaction decomposition, which is suitable for low energies, the free fields obey the following equations:

- The fermion field ψ satisfies the Dirac equation; $\partial - m_f c)\psi_f = 0$ for each type f of fermion.
- The photon field A satisfies the wave equation $\partial_\mu \partial^\mu A^\nu = 0$.
- The Higgs field φ satisfies the Klein–Gordon equation.
- The weak interaction fields Z, $W^\pm$ also satisfy the Proca equation.

These equations can be solved exactly. One usually does so by considering first solutions that are periodic with some period L along each spatial axis; later taking the limit: $L \to \infty$ will lift this periodicity restriction.

In the periodic case, the solution for a field F (any of the above) can be expressed as a Fourier series of the form

$$F(x) = \beta \sum_{\mathbf{p}} \sum_{r} E_{\mathbf{p}}^{-\frac{1}{2}} \left(a_r(\mathbf{p}) u_r(\mathbf{p}) e^{-\frac{ipx}{\hbar}} + b_r^\dagger(\mathbf{p}) v_r(\mathbf{p}) e^{\frac{ipx}{\hbar}} \right)$$

where:

- β is a normalization factor; for the fermion field ψ_f it is $\sqrt{m_f c^2 / V}$, , where

$V = L^3$ is the volume of the fundamental cell considered; for the photon field A^μ it is $\hbar c / \sqrt{2V}$.

- The sum over p is over all momenta consistent with the period L, i.e., over all vectors $\frac{2\pi\hbar}{L}(n_1, n_2, n_3)$ where n_1, n_2, n_3 are integers.
- The sum over r covers other degrees of freedom specific for the field, such as polarization or spin; it usually comes out as a sum from 1 to 2 or from 1 to 3.
- E_p is the relativistic energy for a momentum p quantum of the field, $= \sqrt{m^2c^4 + c^2\mathbf{p}^2}$ when the rest mass is m.
- $a_r(\mathrm{p})$ and $b_r^\dagger(\mathbf{p})$ are annihilation and creation respectively operators for "a-particles" and "b-particles" respectively of momentum p; "b-particles" are the antiparticles of "a-particles". Different fields have different "a-" and "b-particles". For some fields, a and b are the same.
- $u_r(\mathrm{p})$ and $\upsilon_r(\mathrm{p})$ are non-operators which carry the vector or spinor aspects of the field (where relevant).
- $p = (E_\mathrm{p} / c, \mathbf{p})$ is the four-momentum for a quantum with momentum p. $px = p_\mu x^\mu$ denotes an inner product of four-vectors.

In the limit $L \to \infty$, the sum would turn into an integral with help from the V hidden inside β. The numeric value of β also depends on the normalization chosen for $u_r(\mathbf{p})$ and $v_r(\mathbf{p})$.

Technically, $a_r^\dagger(\mathbf{p})$ is the Hermitian adjoint of the operator $a_r(\mathrm{p})$ in the inner product space of ket vectors. The identification of $a_r^\dagger(\mathbf{p})$ and $a_r(\mathrm{p})$ as creation and annihilation operators comes from comparing conserved quantities for a state before and after one of these have acted upon it. can for example be seen to add one particle, because it will add 1 to the eigenvalue of the a-particle number operator, and the momentum of that particle ought to be p since the eigenvalue of the vector-valued momentum operator increases by that much. For these derivations, one starts out with expressions for the operators in terms of the quantum fields. That the operators with † are creation operators and the one without annihilation operators is a convention, imposed by the sign of the commutation relations postulated for them.

An important step in preparation for calculating in perturbative quantum field theory is to separate the "operator" factors a and b above from their corresponding vector or spinor factors u and υ. The vertices of Feynman graphs come from the way that u and υ from different factors in the interaction Lagrangian fit together, whereas the edges come from the way that the as and bs must be moved around in order to put terms in the Dyson series on normal form.

Interaction terms and the path integral approach

The Lagrangian can also be derived without using creation and annihilation operators (the "canonical" formalism), by using a "path integral" approach, pioneered by Feynman building on the earlier work of Dirac. See e.g. A. Zee's QFT in a nutshell. This is one possible way that the Feynman diagrams, which are pictorial representations of interaction terms, can be derived relatively easily. A quick derivation is indeed presented at the article on Feynman diagrams.

Lagrangian Formalism

We can now give some more detail about the aforementioned free and interaction terms appearing in the Standard Model Lagrangian density. Any such term must be both gauge and reference-frame invariant, otherwise the laws of physics would depend on an arbitrary choice or the frame of an observer. Therefore, the global Poincaré symmetry, consisting of translational symmetry, rotational symmetry and the inertial reference frame invariance central to the theory of special relativity must apply. The local $SU(3) \times SU(2) \times U(1)$ gauge symmetry is the internal symmetry. The three factors of the gauge symmetry together give rise to the three fundamental interactions, after some appropriate relations have been defined, as we shall see.

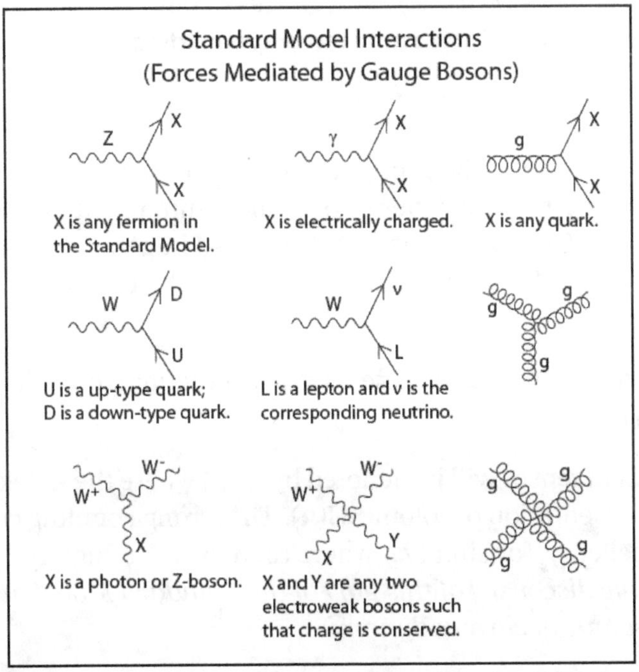

The above interactions show some basic interaction vertices – Feynman diagrams in the standard model are built from these vertices. Higgs boson interactions are however not shown, and neutrino oscillations are commonly added. The charge of the W bosons are dictated by the fermions they interact with.

A complete formulation of the Standard Model Lagrangian with all the terms written together can be found e.g. here.

Kinetic Terms

A free particle can be represented by a mass term, and a *kinetic* term which relates to the "motion" of the fields.

Fermion Fields

The kinetic term for a Dirac fermion is

$$i\bar{\psi}\gamma^{\mu}\partial_{\mu}\psi$$

where the notations are carried from earlier in the article. ψ can represent any, or all, Dirac fermions in the standard model. Generally, as below, this term is included within the couplings (creating an overall "dynamical" term).

Gauge Fields

For the spin-1 fields, first define the field strength tensor

$$F^{a}_{\mu\nu} = \partial_{\mu}A^{a}_{\nu} - \partial_{\nu}A^{a}_{\mu} + gf^{abc}A^{b}_{\mu}A^{c}_{\nu}$$

for a given gauge field (here we use A), with gauge coupling constant g. The quantity f^{abc} is the structure constant of the particular gauge group, defined by the commutator

$$[t_a, t_b] = if^{abc}t_c,$$

where t_i are the generators of the group. In an Abelian (commutative) group (such as the U(1) we use here), since the generators t_a all commute with each other, the structure constants vanish. Of course, this is not the case in general – the standard model includes the non-Abelian SU(2) and SU(3) groups (such groups lead to what is called a Yang–Mills gauge theory).

We need to introduce three gauge fields corresponding to each of the subgroups $SU(3) \times SU(2) \times U(1)$.

- The gluon field tensor will be denoted by $G^{a}_{\mu\nu}$, where the index a labels elements of the 8 representation of colour SU(3). The strong coupling constant is conventionally labelled g_s (or simply g where there is no ambiguity). *The observations leading to the discovery of this part of the Standard Model are discussed in the article in quantum chromodynamics.*
- The notation $W^{a}_{\mu\nu}$ will be used for the gauge field tensor of SU(2) where a runs over the 3 generators of this group. The coupling can be denoted g_w or again simply g. The gauge field will be denoted by W^{a}_{μ}.
- The gauge field tensor for the U(1) of weak hypercharge will be denoted by $B_{\mu\nu}$, the coupling by g', and the gauge field by B_{μ}.

The kinetic term can now be written simply as,

$$\mathcal{L}_{\text{kin}} = -\frac{1}{4}B_{\mu\nu}B^{\mu\nu} - \frac{1}{2}\text{tr}W_{\mu\nu}W^{\mu\nu} - \frac{1}{2}\text{tr}G_{\mu\nu}G^{\mu\nu}$$

where the traces are over the SU(2) and SU(3) indices hidden in W and G respectively. The two-index objects are the field strengths derived from W and G the vector fields. There are also two extra hidden parameters: the theta angles for SU(2) and SU(3).

Coupling Terms

The next step is to "couple" the gauge fields to the fermions, allowing for interactions.

Electroweak Sector

The electroweak sector interacts with the symmetry group U(1) × SU(2)$_L$, where the subscript L indicates coupling only to left-handed fermions.

$$\mathcal{L}_{\text{EW}} = \sum_{\psi} \bar{\psi}\gamma^{\mu}\left(i\partial_{\mu} - g'\frac{1}{2}Y_{\text{W}}B_{\mu} - g\frac{1}{2}\tau\mathbf{W}_{\mu}\right)\psi$$

Where B_μ is the U(1) gauge field; Y_W is the weak hypercharge (the generator of the U(1) group); W_μ is the three-component SU(2) gauge field; and the components of τ are the Pauli matrices (infinitesimal generators of the SU(2) group) whose eigenvalues give the weak isospin. Note that we have to redefine a new U(1) symmetry of *weak hypercharge*, different from QED, in order to achieve the unification with the weak force. The electric charge Q, third component of weak isospin T_3 (also called T_z, I_3 or I_z) and weak hypercharge Y_W are related by,

$$Q = T_3 + \frac{1}{2}Y_W,$$

or by the alternate convention Q = T_3 + Y_W. The first convention (used in this article) is equivalent to the earlier Gell-Mann–Nishijima formula. We can then define the conserved current for weak isospin as,

$$\mathbf{j}_{\mu} = \frac{1}{2}\bar{\psi}_L\gamma_{\mu}\tau\psi_L$$

and for weak hypercharge as,

$$j_{\mu}^{Y} = 2(j_{\mu}^{em} - j_{\mu}^{3})$$

where j_μ^{em} is the electric current and j_μ^3 the third weak isospin current. As explained above, *these currents mix* to create the physically observed bosons, which also leads to testable relations between the coupling constants.

To explain in a simpler way, we can see the effect of the electroweak interaction by picking out terms from the Lagrangian. We see that the SU(2) symmetry acts on each (left-handed) fermion doublet contained in ψ, for example,

$$-\frac{g}{2}(\bar{\nu}_e\,\bar{e})\tau^+\gamma_\mu(W^-)^\mu\begin{pmatrix}\nu_e\\ e\end{pmatrix}=-\frac{g}{2}\bar{\nu}_e\gamma_\mu(W^-)^\mu e$$

where the particles are understood to be left-handed, and where,

$$\tau^+\equiv\frac{1}{2}(\tau^1+i\tau^2)=\begin{pmatrix}0 & 1\\ 0 & 0\end{pmatrix}$$

This is an interaction corresponding to a "rotation in weak isospin space" or in other words, a *transformation between e_L and ν_{eL} via emission of a* W^- *boson*. The U(1) symmetry, on the other hand, is similar to electromagnetism, but acts on all "*weak hypercharged*" fermions (both left and right handed) via the neutral Z^0, as well as the *charged* fermions via the photon.

Quantum Chromodynamics Sector

The quantum chromodynamics (QCD) sector defines the interactions between quarks and gluons, with SU(3) symmetry, generated by T_a. Since leptons do not interact with gluons, they are not affected by this sector. The Dirac Lagrangian of the quarks coupled to the gluon fields is given by,

$$\mathcal{L}_{\text{QCD}}=i\bar{U}\left(\partial_\mu-ig_sG_\mu^aT^a\right)\gamma^\mu U+i\bar{D}\left(\partial_\mu-ig_sG_\mu^aT^a\right)\gamma^\mu D.$$

where D and U are the Dirac spinors associated with up- and down-type quarks, and other notations are continued from the previous section.

Mass Terms and the Higgs Mechanism

Mass Terms

The mass term arising from the Dirac Lagrangian (for any fermion ψ) is $-m\bar{\psi}\psi$ which is *not* invariant under the electroweak symmetry. This can be seen by writing ψ in terms of left and right handed components (skipping the actual calculation):

$$-m\bar{\psi}\psi=-m(\bar{\psi}_L\psi_R+\bar{\psi}_R\psi_L)$$

i.e. contribution from $\bar{\psi}_L\psi_L$ and $\bar{\psi}_R\psi_R$ terms do not appear. We see that the mass-generating interaction is achieved by constant flipping of particle chirality. The spin-half particles have no right/left chirality pair with the same SU(2) representations and equal and opposite weak hypercharges, so assuming these gauge charges are conserved in the vacuum, none of the spin-half particles could ever swap chirality, and must remain massless. Additionally, we know experimentally that the W and Z bosons are massive, but a boson mass term contains the combination e.g. $A^{\mu}A_{\mu}$, which clearly depends on the choice of gauge. Therefore, none of the standard model fermions *or* bosons can "begin" with mass, but must acquire it by some other mechanism.

The Higgs Mechanism

The solution to both these problems comes from the Higgs mechanism, which involves scalar fields (the number of which depend on the exact form of Higgs mechanism) which (to give the briefest possible description) are "absorbed" by the massive bosons as degrees of freedom, and which couple to the fermions via Yukawa coupling to create what looks like mass terms.

In the Standard Model, the Higgs field is a complex scalar of the group $SU(2)_L$:

$$\phi = \frac{1}{\sqrt{2}}\begin{pmatrix}\phi^{+}\\ \phi^{0}\end{pmatrix},$$

where the superscripts + and 0 indicate the electric charge (Q) of the components. The weak hypercharge (Y_W) of both components is 1.

The Higgs part of the Lagrangian is,

$$\mathcal{L}_H = \left[\left(\partial_\mu - igW_\mu^a t^a - ig'Y_\phi B_\mu\right)\phi\right]^2 + \mu^2\phi^\dagger\phi - \lambda(\phi^\dagger\phi)^2,$$

where $\lambda > 0$ and $\mu^2 > 0$, so that the mechanism of spontaneous symmetry breaking can be used. There is a parameter here, at first hidden within the shape of the potential, that is very important. In a unitarity gauge one can set $\varphi^+ = 0$ and make φ^0 real. Then $\phi^0 = v$ is the non-vanishing vacuum expectation value of the Higgs field. v has units of mass, and it is the only parameter in the Standard Model which is not dimensionless. It is also much smaller than the Planck scale; it is approximately equal to the Higgs mass, and sets the scale for the mass of everything else. This is the only real fine-tuning to a small nonzero value in the Standard Model, and it is called the Hierarchy problem. Quadratic terms in W_μ and B_μ arise, which give masses to the W and Z bosons:

$$M_W = \frac{1}{2}v|g|$$

$$M_Z = \frac{1}{2}v\sqrt{g^2 + g'^2}$$

The Yukawa interaction terms are,

$$\mathcal{L}_{YU} = \bar{U}_L G_u U_R \phi^0 - \bar{D}_L G_u U_R \phi^- + \bar{U}_L G_d D_R \phi^+ + \bar{D}_L G_d D_R \phi^0 + hc$$

where $G_{u,d}$ are 3×3 matrices of Yukawa couplings, with the *ij* term giving the coupling of the generations *i* and *j*.

Neutrino Masses

As previously mentioned, evidence shows neutrinos must have mass. But within the standard model, the right-handed neutrino does not exist, so even with a Yukawa coupling neutrinos remain massless. An obvious solution is to simply *add a right-handed neutrino* ν_R resulting in a Dirac mass term as usual. This field however must be a sterile neutrino, since being right-handed it experimentally belongs to an isospin singlet (T_3 = 0) and also has charge Q = 0, implying Y_W = 0 i.e. it does not even participate in the weak interaction. Current experimental status is that evidence for observation of sterile neutrinos is not convincing.

Another possibility to consider is that the neutrino satisfies the Majorana equation, which at first seems possible due to its zero electric charge. In this case the mass term is,

$$-\frac{m}{2}\left(\bar{\nu}^C \nu + \bar{\nu}\nu^C\right)$$

where C denotes a charge conjugated (i.e. anti-) particle, and the terms are consistently all left (or all right) chirality (note that a left-chirality projection of an antiparticle is a right-handed field; care must be taken here due to different notations sometimes used). Here we are essentially flipping between LH neutrinos and RH anti-neutrinos (it is furthermore possible but *not* necessary that neutrinos are their own antiparticle, so these particles are the same). However, for the left-chirality neutrinos, this term changes weak hypercharge by 2 units - not possible with the standard Higgs interaction, requiring the Higgs field to be extended to include an extra triplet with weak hypercharge 2 - whereas for right-chirality neutrinos, no Higgs extensions are necessary. For both left and right chirality cases, Majorana terms violate lepton number, but possibly at a level beyond the current sensitivity of experiments to detect such violations.

It is possible to include both Dirac and Majorana mass terms in the same theory, which (in contrast to the Dirac-mass-only approach) can provide a "natural" explanation for the smallness of the observed neutrino masses, by linking the RH neutrinos to yet-unknown physics around the GUT scale.

Since in any case new fields must be postulated to explain the experimental results, neutrinos are an obvious gateway to searching physics beyond the Standard Model.

Detailed Information

This section provides more detail on some aspects, and some reference material.

Field Content in Detail

The Standard Model has the following fields. These describe one *generation* of leptons and quarks, and there are three generations, so there are three copies of each field. By CPT symmetry, there is a set of right-handed fermions with the opposite quantum numbers. The column "representation" indicates under which representations of the gauge groups that each field transforms, in the order (SU(3), SU(2), U(1)). Symbols used are common but not universal; superscript C denotes an antiparticle; and for the U(1) group, the value of the weak hypercharge is listed. Note that there are twice as many left-handed lepton field components as left-handed antilepton field components in each generation, but an equal number of left-handed quark and antiquark fields.

Free Parameters

Upon writing the most general Lagrangian without neutrinos, one finds that the dynamics depend on 19 parameters, whose numerical values are established by experiment. With neutrinos 7 more parameters are needed, 3 masses and 4 PMNS matrix parameters, for a total of 26 parameters. The neutrino parameter values are still uncertain. The 19 certain parameters are summarized here.

Parameters of the Standard Model			
Symbol	**Description**	**Renormalization scheme (point)**	**Value**
m_e	Electron mass		511 keV
m_μ	Muon mass		105.7 MeV
m_τ	Tau mass		1.78 GeV
m_u	Up quark mass	$\mu_{MS} = 2$ GeV	1.9 MeV
m_d	Down quark mass	$\mu_{MS} = 2$ GeV	4.4 MeV
m_s	Strange quark mass	$\mu_{MS} = 2$ GeV	87 MeV
m_c	Charm quark mass	$\mu_{MS} = m_c$	1.32 GeV
m_b	Bottom quark mass	$\mu_{MS} = m_b$	4.24 GeV
m_t	Top quark mass	On-shell scheme	172.7 GeV
θ_{12}	CKM 12-mixing angle		13.1°
θ_{23}	CKM 23-mixing angle		2.4°
θ_{13}	CKM 13-mixing angle		0.2°
δ	CKM CP-violating Phase		0.995
g_1 or g'	U(1) gauge coupling	$\mu_{MS} = m_Z$	0.357
g_2 or g	SU(2) gauge coupling	$\mu_{MS} = m_Z$	0.652
g_3 or g_s	SU(3) gauge coupling	$\mu_{MS} = m_Z$	1.221

θ_{QCD}	QCD vacuum angle		~0
v	Higgs vacuum expectation value		246 GeV
m_H	Higgs mass		~ 125 GeV (tentative)

The choice of free parameters is somewhat arbitrary. In the table above, gauge couplings are listed as free parameters, therefore with this choice Weinberg angle is not a free parameter - it is defined as $\tan\theta_W = \frac{g_1}{g_2}$. . Likewise, fine structure constant of QED is $\alpha = \frac{1}{4\pi}\frac{(g_1 g_2)^2}{g_1^2 + g_2^2}$.

Instead of fermion masses, dimensionless Yukawa couplings can be chosen as free parameters. For example, electron mass depends on the Yukawa coupling of electron to Higgs field, and its value is $m_e = \frac{y_e v}{\sqrt{2}}$.

Instead of the Higgs mass, the Higgs self-coupling strength $\lambda \sim 1/8$ can be chosen as a free parameter.

Additional Symmetries of the Standard Model

From the theoretical point of view, the Standard Model exhibits four additional global symmetries, not postulated at the outset of its construction, collectively denoted accidental symmetries, which are continuous U(1) global symmetries. The transformations leaving the Lagrangian invariant are:

$$\psi_q(x) \to e^{i\alpha/3}\psi_q$$

$$E_L \to e^{i\beta}E_L \text{ and } (e_R)^c \to e^{i\beta}(e_R)^c$$

$$M_L \to e^{i\beta}M_L \text{ and } (\mu_R)^c \to e^{i\beta}(\mu_R)^c$$

$$T_L \to e^{i\beta}T_L \text{ and } (\tau_R)^c \to e^{i\beta}(\tau_R)^c$$

The first transformation rule is shorthand meaning that all quark fields for all generations must be rotated by an identical phase simultaneously. The fields M_L, T_L and $(\mu_R)^c, (\tau_R)^c$ are the 2nd (muon) and 3rd (tau) generation analogs of E_L and $(e_R)^c$ fields.

By Noether's theorem, each symmetry above has an associated conservation law: the conservation of baryon number, electron number, muon number, and tau number. Each quark is assigned a baryon number of $\frac{1}{3}$, while each antiquark is assigned a baryon number of $-\frac{1}{3}$. Conservation of baryon number implies that the number of quarks minus the number of antiquarks is a constant. Within experimental limits, no violation of this conservation law has been found.

Similarly, each electron and its associated neutrino is assigned an electron number of +1, while the anti-electron and the associated anti-neutrino carry a −1 electron number.

Similarly, the muons and their neutrinos are assigned a muon number of +1 and the tau leptons are assigned a tau lepton number of +1. The Standard Model predicts that each of these three numbers should be conserved separately in a manner similar to the way baryon number is conserved. These numbers are collectively known as lepton family numbers (LF).

In addition to the accidental (but exact) symmetries described above, the Standard Model exhibits several approximate symmetries. These are the "SU(2) custodial symmetry" and the "SU(2) or SU(3) quark flavor symmetry."

Symmetries of the Standard Model and Associated Conservation Laws			
Symmetry	**Lie Group**	**Symmetry Type**	**Conservation Law**
Poincaré	Translations×SO(3,1)	Global symmetry	Energy, Momentum, Angular momentum
Gauge	SU(3)×SU(2)×U(1)	Local symmetry	Color charge, Weak isospin, Electric charge, Weak hypercharge
Baryon phase	U(1)	Accidental Global symmetry	Baryon number
Electron phase	U(1)	Accidental Global symmetry	Electron number
Muon phase	U(1)	Accidental Global symmetry	Muon number
Tau phase	U(1)	Accidental Global symmetry	Tau number

The U(1) Symmetry

For the leptons, the gauge group can be written $SU(2)_l \times U(1)_L \times U(1)_R$. The two U(1) factors can be combined into $U(1)_Y \times U(1)_l$ where l is the lepton number. Gauging of the lepton number is ruled out by experiment, leaving only the possible gauge group $SU(2)_L \times U(1)_Y$. A similar argument in the quark sector also gives the same result for the electroweak theory.

The Charged and Neutral Current Couplings and Fermi Theory

The charged currents $j^{\pm} = j^1 \pm ij^2$ are,

$$j_\mu^+ = \overline{U}_{iL}\gamma_\mu D_{iL} + \overline{v}_{iL}\gamma_\mu l_{iL}.$$

These charged currents are precisely those that entered the Fermi theory of beta decay. The action contains the charge current piece,

$$\mathcal{L}_{CC} = \frac{g}{\sqrt{2}}(j_\mu^+ W^{-\mu} + j_\mu^- W^{+\mu}).$$

For energy much less than the mass of the W-boson, the effective theory becomes the current–current interaction of the Fermi theory.

However, gauge invariance now requires that the component W^3 of the gauge field also be coupled to a current that lies in the triplet of SU(2). However, this mixes with the U(1), and another current in that sector is needed. These currents must be uncharged in order to conserve charge. So we require the neutral currents,

$$j_\mu^3 = \frac{1}{2}(\bar{U}_{iL}\gamma_\mu U_{iL} - \bar{D}_{iL}\gamma_\mu D_{iL} + \bar{\nu}_{iL}\gamma_\mu \nu_{iL} - \bar{l}_{iL}\gamma_\mu l_{iL})$$

$$j_\mu^{em} = \frac{2}{3}\bar{U}_i\gamma_\mu U_i - \frac{1}{3}\bar{D}_i\gamma_\mu D_i - \bar{l}_i\gamma_\mu l_i.$$

The neutral current piece in the Lagrangian is then,

$$\mathcal{L}_{NC} = ej_\mu^{em}A^\mu + \frac{g}{\cos\theta_W}(J_\mu^3 - \sin^2\theta_W J_\mu^{em})Z^\mu.$$

Higgs Mechanism

In the Standard Model of particle physics, the Higgs mechanism is essential to explain the generation mechanism of the property "mass" for gauge bosons. Without the Higgs mechanism, or some other effect like it, all bosons (a type of fundamental particle) would be massless, but measurements show that the W^+, W^-, and Z bosons actually have relatively large masses of around 80 GeV/c^2. The Higgs field resolves this conundrum. The simplest description of the mechanism adds a quantum field (the Higgs field) that permeates all space, to the Standard Model. Below some extremely high temperature, the field causes spontaneous symmetry breaking during interactions. The breaking of symmetry triggers the Higgs mechanism, causing the bosons it interacts with to have mass. In the Standard Model, the phrase "Higgs mechanism" refers specifically to the generation of masses for the $W^\pm$, and Z weak gauge bosons through electroweak symmetry breaking. The Large Hadron Collider at CERN announced results consistent with the Higgs particle on March 14, 2013, making it extremely likely that the field, or one like it, exists, and explaining how the Higgs mechanism takes place in nature.

The mechanism was proposed in 1962 by Philip Warren Anderson, following work in the late 1950s on symmetry breaking in superconductivity and a 1960 paper by Yoichiro Nambu that discussed its application within particle physics. A theory able to finally explain mass generation without "breaking" gauge theory was published almost simultaneously by three independent groups in 1964: by Robert Brout and François Englert; by Peter Higgs; and by Gerald Guralnik, C. R. Hagen, and Tom Kibble. The Higgs mechanism is therefore also called the Brout–Englert–Higgs mechanism or Englert–Brout–Higgs–Guralnik–Hagen–Kibble mechanism, Anderson–Higgs mechanism, Anderson–Higgs-Kibble mechanism, Higgs–Kibble mechanism by Abdus Salam and ABEGHHK'tH mechanism [for Anderson, Brout, Englert, Guralnik, Hagen, Higgs, Kibble and 't Hooft] by Peter Higgs.

On October 8, 2013, following the discovery at CERN's Large Hadron Collider of a new particle that appeared to be the long-sought Higgs boson predicted by the theory, it was announced that Peter Higgs and François Englert had been awarded the 2013 Nobel Prize in Physics (Englert's co-author Robert Brout had died in 2011 and the Nobel Prize is not usually awarded posthumously).

Standard Model

The Higgs mechanism was incorporated into modern particle physics by Steven Weinberg and Abdus Salam, and is an essential part of the standard model.

In the standard model, at temperatures high enough that electroweak symmetry is unbroken, all elementary particles are massless. At a critical temperature, the Higgs field becomes tachyonic; the symmetry is spontaneously broken by condensation, and the W and Z bosons acquire masses. (This also known as electroweak symmetry breaking; *EWSB.*)

Fermions, such as the leptons and quarks in the Standard Model, can also acquire mass as a result of their interaction with the Higgs field, but not in the same way as the gauge bosons.

Structure of the Higgs Field

In the standard model, the Higgs field is a SU(2) doublet (i.e. the standard representation with with two complex components called isospin), which is a scalar under Lorentz transformations. Its (weak hypercharge) U(1) charge is 1. Under U(1) rotations, it is multiplied by a phase, which thus mixes the real and imaginary parts of the complex spinor into each other—combining to the standard two component complex representation of the group U(2).

The Higgs field, through the interactions specified (summarized, represented, or even simulated) by its potential, induces spontaneous breaking of three out of the four generators ("directions") of the gauge group U(2). This is often written asSU(2) × U(1), (which is strictly speaking only the same on the level of infinitessimal symmetries) because the diagonal phase factor also acts on other fields in particular quarks. Three out of its four components would ordinarily amount to Goldstone bosons, if they were not coupled to gauge fields.

However, after symmetry breaking, these three of the four degrees of freedom in the Higgs field mix with the three W and Z bosons (W+, W− and Z), and are only observable as components of these weak bosons, which are now massive; while the one remaining degree of freedom becomes the Higgs boson—a new scalar particle.

The Photon as the Part that Remains Massless

The gauge group of the electroweak part of the standard model is SU(2) × U(1). The group SU(2) is the group of all 2-by-2 unitary matrices with unit determinant; all the orthonormal changes of coordinates in a complex two dimensional vector space.

Rotating the coordinates so that the second basis vector points in the direction of the Higgs boson makes the vacuum expectation value of H the spinor (0, v). The generators for rotations about the x, y, and z axes are by half the Pauli matrices σ_x, σ_y, and σ_z, so that a rotation of angle θ about the z-axis takes the vacuum to,

$$\left(0, ve^{-\frac{1}{2}i\theta}\right).$$

While the T_x and T_y generators mix up the top and bottom components of the spinor, the T_z rotations only multiply each by opposite phases. This phase can be undone by a U(1) rotation of angle 1/2θ. Consequently, under both an SU(2) T_z-rotation and a U(1) rotation by an amount 1/2θ, *the vacuum is invariant.*

This combination of generators,

$$Q = T_z + \frac{Y}{2}$$

defines the unbroken part of the gauge group, where Q is the electric charge, T_z is the generator of rotations around the z-axis in the SU(2) and Y is the hypercharge generator of the U(1). This combination of generators (a z rotation in the SU(2) and a simultaneous U(1) rotation by half the angle) preserves the vacuum, and defines the unbroken gauge group in the standard model, namely *the electric charge* group. The part of the gauge field in this direction stays massless, and amounts to the physical photon.

Consequences for Fermions

In spite of the introduction of spontaneous symmetry breaking, the mass terms oppose the chiral gauge invariance. For these fields the mass terms should always be replaced by a gauge-invariant "Higgs" mechanism. One possibility is some kind of "Yukawa coupling" between the fermion field ψ and the Higgs field Φ, with unknown couplings G_ψ, which after symmetry breaking (more precisely: after expansion of the Lagrange density around a suitable ground state) again results in the original mass terms, which are now, however (i.e. by introduction of the Higgs field) written in a gauge-invariant way. The Lagrange density for the "Yukawa" interaction of a fermion field ψ and the Higgs field Φ is,

$$\mathcal{L}_{\text{Fermion}}(\phi, A, \psi) = \bar{\psi}\gamma^{\mu}D_{\mu}\psi + G_{\psi}\bar{\psi}\phi\psi,$$

where again the gauge field A only enters D_μ (i.e., it is only indirectly visible). The quantities γ^μ are the Dirac matrices, and G_ψ is the already-mentioned "Yukawa" coupling parameter. Already now the mass-generation follows the same principle as above, namely from the existence of a finite expectation value $|\langle\phi\rangle|$, as described above. Again, this is crucial for the existence of the property "mass".

History of Research

Background

Spontaneous symmetry breaking offered a framework to introduce bosons into relativistic quantum field theories. However, according to Goldstone's theorem, these bosons should be massless. The only observed particles which could be approximately interpreted as Goldstone bosons were the pions, which Yoichiro Nambu related to chiral symmetry breaking.

A similar problem arises with Yang–Mills theory (also known as non-abelian gauge theory), which predicts massless spin-1 gauge bosons. Massless weakly interacting gauge bosons lead to long-range forces, which are only observed for electromagnetism and the corresponding massless photon. Gauge theories of the weak force needed a way to describe massive gauge bosons in order to be consistent.

Discovery

The mechanism was proposed in 1962 by Philip Warren Anderson, who discussed its consequences for particle physics but did not work out an explicit relativistic model. The relativistic model was developed in 1964 by three independent groups – Robert Brout and François Englert; Peter Higgs; and Gerald Guralnik, Carl Richard Hagen, and Tom Kibble. Slightly later, in 1965, but independently from the other publications the mechanism was also proposed by Alexander Migdal and Alexander Polyakov, at that time Soviet undergraduate students. However, the paper was delayed by the Editorial Office of JETP, and was published only in 1966.

Number six: Peter Higgs 2009.

Philip W. Anderson, the first to propose the mechanism in 1962.

Five of the six 2010 APS Sakurai Prize Winners – (L to R) Tom Kibble, Gerald Guralnik, Carl Richard Hagen, François Englert, and Robert Brout.

The mechanism is closely analogous to phenomena previously discovered by Yoichiro Nambu involving the "vacuum structure" of quantum fields in superconductivity. A similar but distinct effect (involving an affine realization of what is now recognized as the Higgs field), known as the Stueckelberg mechanism, had previously been studied by Ernst Stueckelberg.

These physicists discovered that when a gauge theory is combined with an additional field that spontaneously breaks the symmetry group, the gauge bosons can consistently acquire a nonzero mass. In spite of the large values involved this permits a gauge theory description of the weak force, which was independently developed by Steven Weinberg and Abdus Salam in 1967. Higgs's original article presenting the model was rejected by Physics Letters. When revising the article before resubmitting it to Physical Review Letters, he added a sentence at the end, mentioning that it implies the existence of one or more new, massive scalar bosons, which do not form complete representations of the symmetry group; these are the Higgs bosons.

The three papers by Brout and Englert; Higgs; and Guralnik, Hagen, and Kibble were each recognized as "milestone letters" by *Physical Review Letters* in 2008. While each of these seminal papers took similar approaches, the contributions and differences among the 1964 PRL symmetry breaking papers are noteworthy. All six physicists were jointly awarded the 2010 J. J. Sakurai Prize for Theoretical Particle Physics for this work.

Benjamin W. Lee is often credited with first naming the "Higgs-like" mechanism, although there is debate around when this first occurred. One of the first times the *Higgs* name appeared in print was in 1972 when Gerardus 't Hooft and Martinus J. G. Veltman referred to it as the "Higgs–Kibble mechanism" in their Nobel winning paper.

Examples

The Higgs mechanism occurs whenever a charged field has a vacuum expectation value. In the nonrelativistic context, this is the Landau model of a charged Bose–Einstein condensate, also known as a superconductor. In the relativistic condensate, the condensate is a scalar field, and is relativistically invariant.

Landau Model

The Higgs mechanism is a type of superconductivity which occurs in the vacuum. It occurs when all of space is filled with a sea of particles which are charged, or, in field language, when a charged field has a nonzero vacuum expectation value. Interaction with the quantum fluid filling the space prevents certain forces from propagating over long distances (as it does in a superconducting medium; e.g., in the Ginzburg–Landau theory).

A superconductor expels all magnetic fields from its interior, a phenomenon known as the Meissner effect. This was mysterious for a long time, because it implies that electromagnetic forces somehow become short-range inside the superconductor. Contrast this with the behavior of an ordinary metal. In a metal, the conductivity shields electric fields by rearranging charges on the surface until the total field cancels in the interior. But magnetic fields can penetrate to any distance, and if a magnetic monopole (an isolated magnetic pole) is surrounded by a metal the field can escape without collimating into a string. In a superconductor, however, electric charges move with no dissipation, and this allows for permanent surface currents, not just surface charges. When magnetic fields are introduced at the boundary of a superconductor, they produce surface currents which exactly neutralize them. The Meissner effect is due to currents in a thin surface layer, whose thickness, the London penetration depth, can be calculated from a simple model (the Ginzburg–Landau theory).

This simple model treats superconductivity as a charged Bose–Einstein condensate. Suppose that a superconductor contains bosons with charge q. The wavefunction of the bosons can be described by introducing a quantum field, ψ, which obeys the

Schrödinger equation as a field equation (in units where the reduced Planck constant, ħ, is set to 1):

$$i\frac{\partial}{\partial t}\psi = \frac{(\nabla - iqA)^2}{2m}\psi.$$

The operator ψ(x) annihilates a boson at the point x, while its adjoint $\psi^\dagger$ creates a new boson at the same point. The wavefunction of the Bose–Einstein condensate is then the expectation value ψ of ψ(x), which is a classical function that obeys the same equation. The interpretation of the expectation value is that it is the phase that one should give to a newly created boson so that it will coherently superpose with all the other bosons already in the condensate.

When there is a charged condensate, the electromagnetic interactions are screened. To see this, consider the effect of a gauge transformation on the field. A gauge transformation rotates the phase of the condensate by an amount which changes from point to point, and shifts the vector potential by a gradient:

$$\begin{aligned} \psi &\to e^{iq\phi(x)}\psi \\ A &\to A + \nabla\phi. \end{aligned}$$

When there is no condensate, this transformation only changes the definition of the phase of ψ at every point. But when there is a condensate, the phase of the condensate defines a preferred choice of phase.

The condensate wave function can be written as,

$$\psi(x) = \rho(x)e^{i\theta(x)},$$

where ρ is real amplitude, which determines the local density of the condensate. If the condensate were neutral, the flow would be along the gradients of θ, the direction in which the phase of the Schrödinger field changes. If the phase θ changes slowly, the flow is slow and has very little energy. But now θ can be made equal to zero just by making a gauge transformation to rotate the phase of the field.

The energy of slow changes of phase can be calculated from the Schrödinger kinetic energy,

$$H = \frac{1}{2m}|(qA + \nabla)\psi|^2,$$

and taking the density of the condensate ρ to be constant,

$$H \approx \frac{\rho^2}{2m}(qA + \nabla\theta)^2.$$

Fixing the choice of gauge so that the condensate has the same phase everywhere, the electromagnetic field energy has an extra term,

$$\frac{q^2\rho^2}{2m}A^2.$$

When this term is present, electromagnetic interactions become short-ranged. Every field mode, no matter how long the wavelength, oscillates with a nonzero frequency. The lowest frequency can be read off from the energy of a long wavelength A mode,

$$E \approx \frac{\dot{A}^2}{2} + \frac{q^2\rho^2}{2m}A^2.$$

This is a harmonic oscillator with frequency,

$$\sqrt{\frac{1}{m}q^2\rho^2}.$$

The quantity $|\psi|^2$ ($=\rho^2$) is the density of the condensate of superconducting particles.

In an actual superconductor, the charged particles are electrons, which are fermions not bosons. So in order to have superconductivity, the electrons need to somehow bind into Cooper pairs. The charge of the condensate q is therefore twice the electron charge e. The pairing in a normal superconductor is due to lattice vibrations, and is in fact very weak; this means that the pairs are very loosely bound. The description of a Bose–Einstein condensate of loosely bound pairs is actually more difficult than the description of a condensate of elementary particles, and was only worked out in 1957 by Bardeen, Cooper and Schrieffer in the famous BCS theory.

Abelian Higgs Mechanism

Gauge invariance means that certain transformations of the gauge field do not change the energy at all. If an arbitrary gradient is added to A, the energy of the field is exactly the same. This makes it difficult to add a mass term, because a mass term tends to push the field toward the value zero. But the zero value of the vector potential is not a gauge invariant idea. What is zero in one gauge is nonzero in another.

So in order to give mass to a gauge theory, the gauge invariance must be broken by a condensate. The condensate will then define a preferred phase, and the phase of the condensate will define the zero value of the field in a gauge-invariant way. The gauge-invariant definition is that a gauge field is zero when the phase change along any path from parallel transport is equal to the phase difference in the condensate wavefunction.

The condensate value is described by a quantum field with an expectation value, just as in the Ginzburg-Landau model.

In order for the phase of the vacuum to define a gauge, the field must have a phase (also referred to as 'to be charged'). In order for a scalar field Φ to have a phase, it must be complex, or (equivalently) it should contain two fields with a symmetry which rotates them into each other. The vector potential changes the phase of the quanta produced by the field when they move from point to point. In terms of fields, it defines how much to rotate the real and imaginary parts of the fields into each other when comparing field values at nearby points.

The only renormalizable model where a complex scalar field Φ acquires a nonzero value is the Mexican-hat model, where the field energy has a minimum away from zero. The action for this model is,

$$S(\phi) = \int \frac{1}{2} |\partial\phi|^2 - \lambda\left(|\phi|^2 - \Phi^2\right)^2,$$

which results in the Hamiltonian,

$$H(\phi) = \frac{1}{2}|\dot{\phi}|^2 + |\nabla\phi|^2 + V(|\phi|).$$

The first term is the kinetic energy of the field. The second term is the extra potential energy when the field varies from point to point. The third term is the potential energy when the field has any given magnitude.

This potential energy, $V(z, \Phi) = \lambda(|z|^2 - \Phi^2)^2$, has a graph which looks like a Mexican hat, which gives the model its name. In particular, the minimum energy value is not at $z = 0$, but on the circle of points where the magnitude of z is Φ.

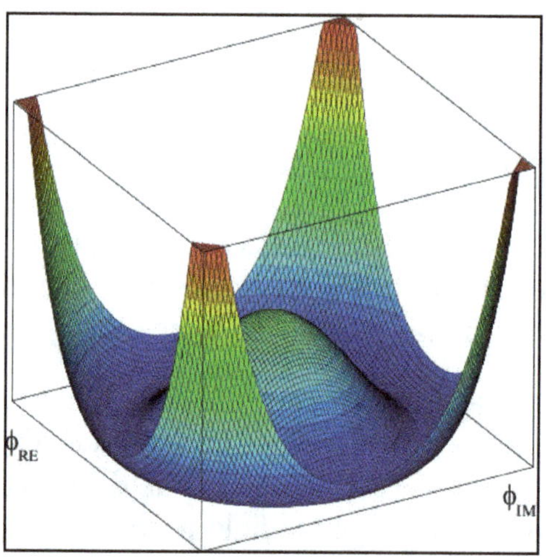

Higgs potential *V*. For a fixed value of λ the potential is presented upwards against the real and imaginary parts of Φ. The *Mexican-hat* or *champagne-bottle profile* at the ground should be noted.

When the field Φ(*x*) is not coupled to electromagnetism, the Mexican-hat potential has flat directions. Starting in any one of the circle of vacua and changing the phase of the field from point to point costs very little energy. Mathematically, if,

$$\phi(x) = \Phi e^{i\theta(x)}$$

with a constant prefactor, then the action for the field $\theta(x)$, i.e., the "phase" of the Higgs field $\Phi(x)$, has only derivative terms. This is not a surprise. Adding a constant to $\theta(x)$ is a symmetry of the original theory, so different values of $\theta(x)$ cannot have different energies. This is an example of Goldstone's theorem: spontaneously broken continuous symmetries normally produce massless excitations.

The Abelian Higgs model is the Mexican-hat model coupled to electromagnetism:

$$S(\phi, A) = \int -\frac{1}{4} F^{\mu\nu} F_{\mu\nu} + |(\partial - iqA)\phi|^2 - \lambda(|\phi|^2 - \Phi^2)^2.$$

The classical vacuum is again at the minimum of the potential, where the magnitude of the complex field φ is equal to Φ. But now the phase of the field is arbitrary, because gauge transformations change it. This means that the field $\theta(x)$ can be set to zero by a gauge transformation, and does not represent any actual degrees of freedom at all.

Furthermore, choosing a gauge where the phase of the vacuum is fixed, the potential energy for fluctuations of the vector field is nonzero. So in the abelian Higgs model, the gauge field acquires a mass. To calculate the magnitude of the mass, consider a constant value of the vector potential A in the x direction in the gauge where the condensate has constant phase. This is the same as a sinusoidally varying condensate in the gauge where the vector potential is zero. In the gauge where A is zero, the potential energy density in the condensate is the scalar gradient energy:

$$E = \frac{1}{2}\left|\partial\left(\Phi e^{iqAx}\right)\right|^2 = \frac{1}{2} q^2 \Phi^2 A^2.$$

This energy is the same as a mass term $1/2m^2A^2$ where $m = q\Phi$.

Nonabelian Higgs Mechanism

The Nonabelian Higgs model has the following action:

$$S(\phi, \mathbf{A}) = \int \frac{1}{4g^2} \mathrm{tr}(F^{\mu\nu} F_{\mu\nu}) + |D\phi|^2 + V(|\phi|)$$

where now the nonabelian field A is contained in the covariant derivative D and in the tensor components $F^{\mu\nu}$ and $F_{\mu\nu}$ (the relation between A and those components is well-known from the Yang–Mills theory).

It is exactly analogous to the Abelian Higgs model. Now the field φ is in a representation of the gauge group, and the gauge covariant derivative is defined by the rate of change of the field minus the rate of change from parallel transport using the gauge field A as a connection.

$$D\phi = \partial\phi - iA^k t_k \phi$$

Again, the expectation value of Φ defines a preferred gauge where the vacuum is constant, and fixing this gauge, fluctuations in the gauge field A come with a nonzero energy cost.

Depending on the representation of the scalar field, not every gauge field acquires a mass. A simple example is in the renormalizable version of an early electroweak model due to Julian Schwinger. In this model, the gauge group is SO(3) (or SU(2) – there are no spinor representations in the model), and the gauge invariance is broken down to U(1) or SO(2) at long distances. To make a consistent renormalizable version using the Higgs mechanism, introduce a scalar field φ^a which transforms as a vector (a triplet) of SO(3). If this field has a vacuum expectation value, it points in some direction in field space. Without loss of generality, one can choose the *z*-axis in field space to be the direction that φ is pointing, and then the vacuum expectation value of φ is (0, 0, *A*), where *A* is a constant with dimensions of mass ($c = \hbar = 1$).

Rotations around the *z*-axis form a U(1) subgroup of SO(3) which preserves the vacuum expectation value of φ, and this is the unbroken gauge group. Rotations around the *x* and *y*-axis do not preserve the vacuum, and the components of the SO(3) gauge field which generate these rotations become massive vector mesons. There are two massive W mesons in the Schwinger model, with a mass set by the mass scale *A*, and one massless U(1) gauge boson, similar to the photon.

The Schwinger model predicts magnetic monopoles at the electroweak unification scale, and does not predict the Z meson. It doesn't break electroweak symmetry properly as in nature. But historically, a model similar to this (but not using the Higgs mechanism) was the first in which the weak force and the electromagnetic force were unified.

Affine Higgs Mechanism

Ernst Stueckelberg discovered a version of the Higgs mechanism by analyzing the theory of quantum electrodynamics with a massive photon. Effectively, Stueckelberg's model is a limit of the regular Mexican hat Abelian Higgs model, where the vacuum expectation value *H* goes to infinity and the charge of the Higgs field goes to zero in such a way that their product stays fixed. The mass of the Higgs boson is proportional to *H*, so the Higgs boson becomes infinitely massive and decouples, so is not present in the discussion. The vector meson mass, however, equals to the product *eH*, and stays finite.

The interpretation is that when a U(1) gauge field does not require quantized charges, it is possible to keep only the angular part of the Higgs oscillations, and discard the radial part. The angular part of the Higgs field θ has the following gauge transformation law:

$$\begin{aligned} \theta &\to \theta + e\alpha \\ A &\to A + \partial\alpha. \end{aligned}$$

The gauge covariant derivative for the angle (which is actually gauge invariant) is:

$$D\theta = \partial\theta - eAH.$$

In order to keep θ fluctuations finite and nonzero in this limit, θ should be rescaled by H, so that its kinetic term in the action stays normalized. The action for the theta field is read off from the Mexican hat action by substituting $\phi = He^{\frac{1}{H}i\theta}$.

$$S = \int \frac{1}{4}F^2 + \frac{1}{2}(D\theta)^2 = \int \frac{1}{4}F^2 + \frac{1}{2}(\partial\theta - HeA)^2 = \int \frac{1}{4}F^2 + \frac{1}{2}(\partial\theta - mA)^2.$$

since *eH* is the gauge boson mass. By making a gauge transformation to set θ = 0, the gauge freedom in the action is eliminated, and the action becomes that of a massive vector field:

$$S = \int \frac{1}{4}F^2 + \frac{1}{2}m^2A^2.$$

To have arbitrarily small charges requires that the U(1) is not the circle of unit complex numbers under multiplication, but the real numbers R under addition, which is only different in the global topology. Such a U(1) group is *non-compact*. The field θ transforms as an affine representation of the gauge group. Among the allowed gauge groups, only non-compact U(1) admits affine representations, and the U(1) of electromagnetism is experimentally known to be compact, since charge quantization holds to extremely high accuracy.

The Higgs condensate in this model has infinitesimal charge, so interactions with the Higgs boson do not violate charge conservation. The theory of quantum electrodynamics with a massive photon is still a renormalizable theory, one in which electric charge is still conserved, but magnetic monopoles are not allowed. For nonabelian gauge theory, there is no affine limit, and the Higgs oscillations cannot be too much more massive than the vectors.

Cabibbo–Kobayashi–Maskawa Matrix

In the Standard Model of particle physics, the Cabibbo–Kobayashi–Maskawa matrix (CKM matrix, quark mixing matrix, sometimes also called KM matrix) is a unitary matrix which contains information on the strength of flavour-changing weak decays. Technically, it specifies the mismatch of quantum states of quarks when they propagate freely and when they take part in the weak interactions. It is important in the understanding of CP violation. This matrix was introduced for three generations of quarks by Makoto Kobayashi and Toshihide Maskawa, adding one generation to the matrix previously introduced by Nicola Cabibbo. This matrix is also an extension of the GIM mechanism, which only includes two of the three current families of quarks.

The Matrix

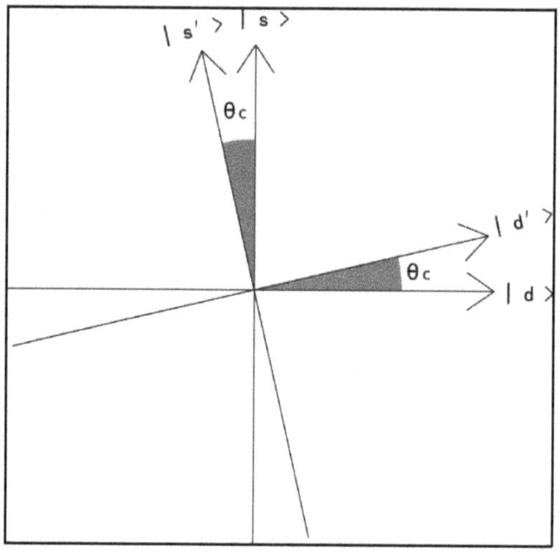

The Cabibbo angle represents the rotation of the mass eigenstate vector space formed by the mass eigenstates into the weak eigenstate vector space formed by the weak eigenstates . $\theta_c = 13.02°$.

In 1963, Nicola Cabibbo introduced the Cabibbo angle (θ_c) to preserve the universality of the weak interaction. Cabibbo was inspired by previous work by Murray Gell-Mann and Maurice Lévy, on the effectively rotated nonstrange and strange vector and axial weak currents, which he references.

In light of current knowledge (quarks were not yet theorized), the Cabibbo angle is related to the relative probability that down and strange quarks decay into up quarks ($|V_{ud}|^2$ and $|V_{us}|^2$ respectively). In particle physics parlance, the object that couples to the up quark via charged-current weak interaction is a superposition of down-type quarks, here denoted by d'. Mathematically this is:

$$d' = V_{ud}d + V_{us}s,$$

or using the Cabibbo angle:

$$d' = \cos\theta_c d + \sin\theta_c s.$$

Using the currently accepted values for $|V_{ud}|$ and $|V_{us}|$, the Cabibbo angle can be calculated using,

$$\tan\theta_c = \frac{|V_{us}|}{|V_{ud}|} = \frac{0.22534}{0.97427} \rightarrow \theta_c = 13.02^\circ.$$

When the charm quark was discovered in 1974, it was noticed that the down and strange quark could decay into either the up or charm quark, leading to two sets of equations:

$$d' = V_{ud}d + V_{us}s;$$

$$s' \quad V_{cd}d \quad V_{cs}s$$

or using the Cabibbo angle:

$$d' = \cos\theta_c d + \sin\theta_c s;$$

$$s' = -\sin\theta_c d + \cos\theta_c s.$$

This can also be written in matrix notation as:

$$\begin{bmatrix} d' \\ s' \end{bmatrix} = \begin{bmatrix} V_{ud} & V_{us} \\ V_{cd} & V_{cs} \end{bmatrix} \begin{bmatrix} d \\ s \end{bmatrix},$$

or using the Cabibbo angle ,

$$\begin{bmatrix} d' \\ s' \end{bmatrix} = \begin{bmatrix} \cos\theta_c & \sin\theta_c \\ -\sin\theta_c & \cos\theta_c \end{bmatrix} \begin{bmatrix} d \\ s \end{bmatrix},$$

where the various $|V_{ij}|^2$ represent the probability that the quark of *j* flavor decays into a quark of *i* flavor. This 2 × 2 rotation matrix is called the Cabibbo matrix.

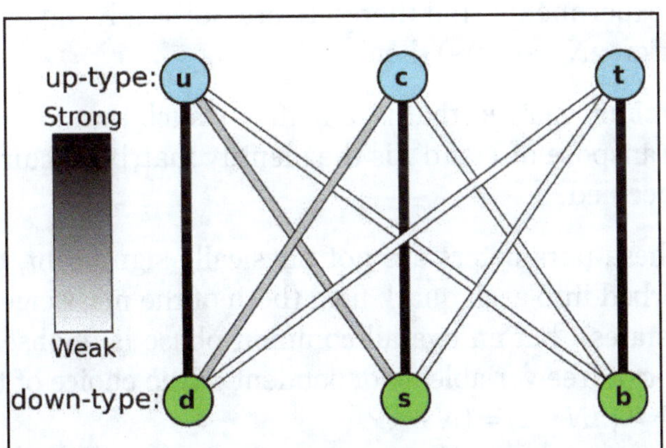

A pictorial representation of the six quarks' decay modes, with mass increasing from left to right.

Observing that CP-violation could not be explained in a four-quark model, Kobayashi and Maskawa generalized the Cabibbo matrix into the Cabibbo–Kobayashi–Maskawa matrix (or CKM matrix) to keep track of the weak decays of three generations of quarks:

$$\begin{bmatrix} d' \\ s' \\ b' \end{bmatrix} = \begin{bmatrix} V_{ud} & V_{us} & V_{ub} \\ V_{cd} & V_{cs} & V_{cb} \\ V_{td} & V_{ts} & V_{tb} \end{bmatrix} \begin{bmatrix} d \\ s \\ b \end{bmatrix}.$$

On the left is the weak interaction doublet partners of up-type quarks, and on the right is the CKM matrix along with a vector of mass eigenstates of down-type quarks. The CKM matrix describes the probability of a transition from one quark i to another quark j. These transitions are proportional to $|V_{ij}|^2$.

Currently, the best determination of the magnitudes of the CKM matrix elements is:

$$\begin{bmatrix} |V_{ud}| & |V_{us}| & |V_{ub}| \\ |V_{cd}| & |V_{cs}| & |V_{cb}| \\ |V_{td}| & |V_{ts}| & |V_{tb}| \end{bmatrix} = \begin{bmatrix} 0.97427 \pm 0.00015 & 0.22534 \pm 0.00065 & 0.00351^{+0.00015}_{-0.00014} \\ 0.22520 \pm 0.00065 & 0.97344 \pm 0.00016 & 0.0412^{+0.0011}_{-0.0005} \\ 0.00867^{+0.00029}_{-0.00031} & 0.0404^{+0.0011}_{-0.0005} & 0.999146^{+0.000021}_{-0.000046} \end{bmatrix}.$$

Note that the choice of usage of down-type quarks in the definition is purely arbitrary and does not represent some sort of deep physical asymmetry between up-type and down-type quarks. We could just as easily define the matrix the other way around, describing weak interaction partners of mass eigenstates of up-type quarks, u', c' and t', in terms of u, c, and t. Since the CKM matrix is unitary (and therefore its inverse is the same as its conjugate transpose), we would obtain essentially the same matrix.

Counting

To proceed further, it is necessary to count the number of parameters in this matrix, V which appear in experiments, and therefore are physically important. If there are N generations of quarks ($2N$ flavours) then

- An $N \times N$ unitary matrix (that is, a matrix V such that $VV^\dagger = I$, where $V^\dagger$ is the conjugate transpose of V and I is the identity matrix) requires N^2 real parameters to be specified.
- $2N - 1$ of these parameters are not physically significant, because one phase can be absorbed into each quark field (both of the mass eigenstates, and of the weak eigenstates), but an overall common phase is unobservable. Hence, the total number of free variables independent of the choice of the phases of basis vectors is $N^2 - (2N - 1) = (N - 1)^2$.
    - Of these, $N(N - 1)/2$ are rotation angles called quark *mixing angles*.
    - The remaining $(N - 1)(N - 2)/2$ are complex phases, which cause CP violation.

For the case $N = 2$, there is only one parameter which is a mixing angle between two generations of quarks. Historically, this was the first version of CKM matrix when only two generations were known. It is called the Cabibbo angle after its inventor Nicola Cabibbo.

For the Standard Model case ($N = 3$), there are three mixing angles and one CP-violating complex phase.

Observations and Predictions

Cabibbo's idea originated from a need to explain two observed phenomena:

1. the transitions $u \leftrightarrow d$, $e \leftrightarrow \nu_e$, and $\mu \leftrightarrow \nu_\mu$ had similar amplitudes.
2. the transitions with change in strangeness $\Delta S = 1$ had amplitudes equal to 1/4 of those with $\Delta S = 0$.

Cabibbo's solution consisted of postulating weak universality to resolve the first issue, along with a mixing angle θ_c, now called the *Cabibbo angle*, between the *d* and *s* quarks to resolve the second.

For two generations of quarks, there are no CP violating phases, as shown by the counting of the previous section. Since CP violations were seen in neutral kaon decays already in 1964, the emergence of the Standard Model soon after was a clear signal of the existence of a third generation of quarks, as pointed out in 1973 by Kobayashi and Maskawa. The discovery of the bottom quark at Fermilab (by Leon Lederman's group) in 1976 therefore immediately started off the search for the missing third-generation quark, the top quark.

Note, however, that the specific values of the angles are *not* a prediction of the standard model: they are open, unfixed parameters. At this time, there is no generally accepted theory that explains why the measured values are what they are.

Weak Universality,

The constraints of unitarity of the CKM-matrix on the diagonal terms can be written as

$$\sum_k |V_{ik}|^2 = \sum_i |V_{ik}|^2 = 1$$

for all generations *i*. This implies that the sum of all couplings of any of the up-type quarks to all the down-type quarks is the same for all generations. This relation is called *weak universality* and was first pointed out by Nicola Cabibbo in 1967. Theoretically it is a consequence of the fact that all SU(2) doublets couple with the same strength to the vector bosons of weak interactions. It has been subjected to continuing experimental tests.

The Unitarity Triangles

The remaining constraints of unitarity of the CKM-matrix can be written in the form,

$$\sum_k V_{ik} V_{jk}^* = 0.$$

For any fixed and different *i* and *j*, this is a constraint on three complex numbers, one for each *k*, which says that these numbers form the sides of a triangle in the complex plane. There are six choices of *i* and *j* (three independent), and hence six such triangles, each of which is called a *unitary triangle*. Their shapes can be very different, but they all have the same area, which can be related to the CP violating phase. The area vanishes for the specific parameters in the Standard Model for which there would be no CP violation. The orientation of the triangles depend on the phases of the quark fields.

Since the three sides of the triangles are open to direct experiment, as are the three angles, a class of tests of the Standard Model is to check that the triangle closes. This is the purpose of a modern series of experiments under way at the Japanese BELLE and the American BaBar experiments, as well as at LHCb in CERN, Switzerland.

Parameterizations

Four independent parameters are required to fully define the CKM matrix. Many parameterizations have been proposed, and three of the most common ones are shown below.

KM Parameters

The original parameterization of Kobayashi and Maskawa used three angles (θ_1, θ_2, θ_3) and a CP-violating phase (δ). Cosines and sines of the angles are denoted c_i and s_i, respectively. θ_1 is the Cabibbo angle.

$$\begin{bmatrix} c_1 & -s_1 c_3 & -s_1 s_3 \\ s_1 c_2 & c_1 c_2 c_3 - s_2 s_3 e^{i\delta} & c_1 c_2 s_3 + s_2 c_3 e^{i\delta} \\ s_1 s_2 & c_1 s_2 c_3 + c_2 s_3 e^{i\delta} & c_1 s_2 s_3 - c_2 c_3 e^{i\delta} \end{bmatrix}.$$

"Standard" Parameters

A "standard" parameterization of the CKM matrix uses three Euler angles (θ_{12}, θ_{23}, θ_{13}) and one CP-violating phase (δ_{13}). Couplings between quark generation *i* and *j* vanish if $\theta_{ij} = 0$. Cosines and sines of the angles are denoted c_{ij} and s_{ij}, respectively. θ_{12} is the Cabibbo angle.

$$\begin{bmatrix} 1 & 0 & 0 \\ 0 & c_{23} & s_{23} \\ 0 & -s_{23} & c_{23} \end{bmatrix} \begin{bmatrix} c_{13} & 0 & s_{13} e^{-i\delta_{13}} \\ 0 & 1 & 0 \\ -s_{13} e^{i\delta_{13}} & 0 & c_{13} \end{bmatrix} \begin{bmatrix} c_{12} & s_{12} & 0 \\ -s_{12} & c_{12} & 0 \\ 0 & 0 & 1 \end{bmatrix}$$

$$= \begin{bmatrix} c_{12}c_{13} & s_{12}c_{13} & s_{13}e^{-i\delta_{13}} \\ -s_{12}c_{23} - c_{12}s_{23}s_{13}e^{i\delta_{13}} & c_{12}c_{23} - s_{12}s_{23}s_{13}e^{i\delta_{13}} & s_{23}c_{13} \\ s_{12}s_{23} - c_{12}c_{23}s_{13}e^{i\delta_{13}} & -c_{12}s_{23} - s_{12}c_{23}s_{13}e^{i\delta_{13}} & c_{23}c_{13} \end{bmatrix}.$$

The currently best known values for the standard parameters are:

$\theta_{12} = 13.04\pm0.05°$, $\theta_{13} = 0.201\pm0.011°$, $\theta_{23} = 2.38\pm0.06°$, and $\delta_{13} = 1.20\pm0.08$ rad.

Wolfenstein Parameters

A third parameterization of the CKM matrix was introduced by Lincoln Wolfenstein with the four parameters λ, A, ρ, and η. The four Wolfenstein parameters have the property that all are of order 1 and are related to the "standard" parameterization:

$\lambda = s_{12}$

$A\lambda^2 = s_{23}$

$A\lambda^3(\rho - i\eta) = s_{13}e^{-i\delta}$

The Wolfenstein parameterization of the CKM matrix, is an approximation of the standard parameterization. To order λ^3, it is:

$$\begin{bmatrix} 1-\lambda^2/2 & \lambda & A\lambda^3(\rho-i\eta) \\ -\lambda & 1-\lambda^2/2 & A\lambda^2 \\ A\lambda^3(1-\rho-i\eta) & -A\lambda^2 & 1 \end{bmatrix}.$$

The CP violation can be determined by measuring $\rho - i\eta$.

Using the values of the previous section for the CKM matrix, the best determination of the Wolfenstein parameters is:

$\lambda = 0.2257^{+0.0009}_{-0.0010}$,

$A = 0.814^{+0.021}_{-0.022}$,

$\rho = 0.135^{+0.031}_{-0.016}$,

and $\eta = 0.349^{+0.015}_{-0.017}$.

Nobel Prize

In 2008, Kobayashi and Maskawa shared one half of the Nobel Prize in Physics "for the discovery of the origin of the broken symmetry which predicts the existence of at least three families of quarks in nature". Some physicists were reported to harbor bitter feelings about the fact that the Nobel Prize committee failed to reward the work of Cabibbo, whose prior work was closely related to that of Kobayashi and Maskawa. Asked for a reaction on the prize, Cabibbo preferred to give no comment.

Spontaneous Symmetry Breaking

Spontaneous symmetry breaking is a mode of realization of symmetry breaking in a physical system, where the underlying laws are invariant under a symmetry transformation, but the system as a whole changes under such transformations, in contrast to explicit symmetry breaking. It is a spontaneous process by which a system in a symmetrical state ends up in an asymmetrical state. It thus describes systems where the equations of motion or the Lagrangian obey certain symmetries, but the lowest-energy solutions do not exhibit that symmetry.

Consider a symmetrical upward dome with a trough circling the bottom. If a ball is put at the very peak of the dome, the system is symmetrical with respect to a rotation around the center axis. But the ball may *spontaneously break* this symmetry by rolling down the dome into the trough, a point of lowest energy. Afterward, the ball has come to a rest at some fixed point on the perimeter. The dome and the ball retain their individual symmetry, but the system does not.

Most simple phases of matter and phase transitions, like crystals, magnets, and conventional superconductors can be simply understood from the viewpoint of spontaneous symmetry breaking. Notable exceptions include topological phases of matter like the fractional quantum Hall effect.

Spontaneous Symmetry Breaking in Physics

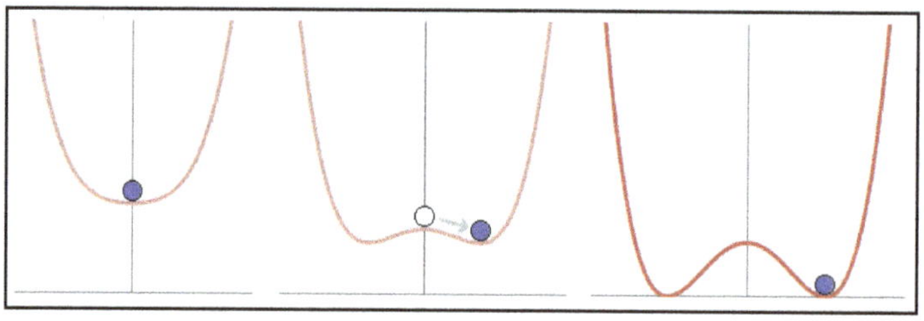

Spontaneous symmetry breaking illustrated: At high energy levels (*left*) the ball settles in the center, and the result is symmetrical. At lower energy levels (*right*), the overall "rules" remain symmetrical, but the symmetric "Mexican hat" enforces an asymmetric outcome, since eventually the ball must rest at some random spot on the bottom, "spontaneously", and not all others.

Particle Physics

In particle physics the force carrier particles are normally specified by field equations with gauge symmetry; their equations predict that certain measurements will be the same at any point in the field. For instance, field equations might predict that the mass of two quarks is constant. Solving the equations to find the mass of each quark might give two solutions. In one solution, quark A is heavier than quark B. In the second solution, quark B is heavier than quark A *by the same amount*. The symmetry of the equations is not reflected by the individual solutions, but it is reflected by the range of solutions.

An actual measurement reflects only one solution, representing a breakdown in the symmetry of the underlying theory. "Hidden" is a better term than "broken", because the symmetry is always there in these equations. This phenomenon is called *spontaneous* symmetry breaking (SSB) because *nothing* (that we know of) breaks the symmetry in the equations.

Chiral Symmetry

Chiral symmetry breaking is an example of spontaneous symmetry breaking affecting the chiral symmetry of the strong interactions in particle physics. It is a property of quantum chromodynamics, the quantum field theory describing these interactions, and is responsible for the bulk of the mass (over 99%) of the nucleons, and thus of all common matter, as it converts very light bound quarks into 100 times heavier constituents of baryons. The approximate Nambu–Goldstone bosons in this spontaneous symmetry breaking process are the pions, whose mass is an order of magnitude lighter than the mass of the nucleons. It served as the prototype and significant ingredient of the Higgs mechanism underlying the electroweak symmetry breaking.

Higgs Mechanism

The strong, weak, and electromagnetic forces can all be understood as arising from gauge symmetries. The Higgs mechanism, the spontaneous symmetry breaking of gauge symmetries, is an important component in understanding the superconductivity of metals and the origin of particle masses in the standard model of particle physics. One important consequence of the distinction between true symmetries and *gauge symmetries*, is that the spontaneous breaking of a gauge symmetry does not give rise to characteristic massless Nambu–Goldstone physical modes, but only massive modes, like the plasma mode in a superconductor, or the Higgs mode observed in particle physics.

In the standard model of particle physics, spontaneous symmetry breaking of the SU(2) × U(1) gauge symmetry associated with the electro-weak force generates masses for several particles, and separates the electromagnetic and weak forces. The W and Z bosons are the elementary particles that mediate the weak interaction, while the photon mediates the electromagnetic interaction. At energies much greater than 100 GeV all these particles behave in a similar manner. The Weinberg–Salam theory predicts that, at lower energies, this symmetry is broken so that the photon and the massive W and Z bosons emerge. In addition, fermions develop mass consistently.

Without spontaneous symmetry breaking, the Standard Model of elementary particle interactions requires the existence of a number of particles. However, some particles (the W and Z bosons) would then be predicted to be massless, when, in reality, they are observed to have mass. To overcome this, spontaneous symmetry breaking is augmented by the Higgs mechanism to give these particles mass. It also suggests the presence of a new particle, the Higgs boson, detected in 2012.

Superconductivity of metals is a condensed-matter analog of the Higgs phenomena, in which a condensate of Cooper pairs of electrons spontaneously breaks the U(1) gauge symmetry associated with light and electromagnetism.

Condensed Matter Physics

Most phases of matter can be understood through the lens of spontaneous symmetry breaking. For example, crystals are periodic arrays of atoms that are not invariant under all translations (only under a small subset of translations by a lattice vector). Magnets have north and south poles that are oriented in a specific direction, breaking rotational symmetry. In addition to these examples, there are a whole host of other symmetry-breaking phases of matter including nematic phases of liquid crystals, charge- and spin-density waves, superfluids and many others.

There are several known examples of matter that cannot be described by spontaneous symmetry breaking, including: topologically ordered phases of matter like fractional quantum Hall liquids, and spin-liquids. These states do not break any symmetry, but are distinct phases of matter. Unlike the case of spontaneous symmetry breaking, there is not a general framework for describing such states.

Continuous Symmetry

The ferromagnet is the canonical system which spontaneously breaks the continuous symmetry of the spins below the Curie temperature and at $h = 0$, where h is the external magnetic field. Below the Curie temperature the energy of the system is invariant under inversion of the magnetization $m(x)$ such that $m(x) = -m(-x)$. The symmetry is spontaneously broken as $h \to 0$ when the Hamiltonian becomes invariant under the inversion transformation, but the expectation value is not invariant.

Spontaneously, symmetry broken phases of matter are characterized by an order parameter that describes the quantity which breaks the symmetry under consideration. For example, in a magnet, the order parameter is the local magnetization.

Spontaneously breaking of a continuous symmetry is inevitably accompanied by gapless (meaning that these modes do not cost any energy to excite) Nambu–Goldstone modes associated with slow long-wavelength fluctuations of the order parameter. For example, vibrational modes in a crystal, known as phonons, are associated with slow density fluctuations of the crystal's atoms. The associated Goldstone mode for magnets are oscillating waves of spin known as spin-waves. For symmetry-breaking states, whose order parameter is not a conserved quantity, Nambu–Goldstone modes are typically massless and propagate at a constant velocity.

An important theorem, due to Mermin and Wagner, states that, at finite temperature, thermally activated fluctuations of Nambu–Goldstone modes destroy the long-range order, and prevent spontaneous symmetry breaking in one- and two-dimensional systems. Similarly, quantum fluctuations of the order parameter prevent most types of continuous symmetry breaking in one-dimensional systems even at zero temperature (an important exception is ferromagnets, whose order parameter, magnetization, is an exactly conserved quantity and does not have any quantum fluctuations).

Other long-range interacting systems such as cylindrical curved surfaces interacting via the Coulomb potential or Yukawa potential has been shown to break translational and rotational symmetries. It was shown, in the presence of a symmetric Hamiltonian, and in the limit of infinite volume, the system spontaneously adopts a chiral configuration, i.e. breaks mirror plane symmetry.

Dynamical Symmetry Breaking

Dynamical symmetry breaking (DSB) is a special form of spontaneous symmetry breaking where the ground state of the system has reduced symmetry properties compared to its theoretical description (Lagrangian).

Dynamical breaking of a global symmetry is a spontaneous symmetry breaking, that happens not at the (classical) tree level (i.e. at the level of the bare action), but due to quantum corrections (i.e. at the level of the effective action).

Dynamical breaking of a gauge symmetry is subtler. In the conventional spontaneous gauge symmetry breaking, there exists an unstable Higgs particle in the theory, which drives the vacuum to a symmetry-broken phase (see e.g. Electroweak interaction). In dynamical gauge symmetry breaking, however, no unstable Higgs particle operates in the theory, but the bound states of the system itself provide the unstable fields that render the phase transition. For example, Bardeen, Hill, and Lindner published a paper which attempts to replace the conventional Higgs mechanism in the standard model, by a DSB that is driven by a bound state of top-antitop quarks (such models, where a composite particle plays the role of the

Higgs boson, are often referred to as "Composite Higgs models"). Dynamical breaking of gauge symmetries is often due to creation of a fermionic condensate; for example the quark condensate, which is connected to the dynamical breaking of chiral symmetry in quantum chromodynamics. Conventional superconductivity is the paradigmatic example from the condensed matter side, where phonon-mediated attractions lead electrons to become bound in pairs and then condense, thereby breaking the electromagnetic gauge symmetry.

Generalisation and Technical Usage

For spontaneous symmetry breaking to occur, there must be a system in which there are several equally likely outcomes. The system as a whole is therefore symmetric with respect to these outcomes. (If we consider any two outcomes, the probability is the same. This contrasts sharply to explicit symmetry breaking.) However, if the system is sampled (i.e. if the system is actually used or interacted with in any way), a specific outcome must occur. Though the system as a whole is symmetric, it is never encountered with this symmetry, but only in one specific asymmetric state. Hence, the symmetry is said to be spontaneously broken in that theory. Nevertheless, the fact that each outcome is equally likely is a reflection of the underlying symmetry, which is thus often dubbed "hidden symmetry", and has crucial formal consequences.

When a theory is symmetric with respect to a symmetry group, but requires that one element of the group be distinct, then spontaneous symmetry breaking has occurred. The theory must not dictate *which* member is distinct, only that *one is*. From this point on, the theory can be treated as if this element actually is distinct, with the proviso that any results found in this way must be resymmetrized, by taking the average of each of the elements of the group being the distinct one.

The crucial concept in physics theories is the order parameter. If there is a field (often a background field) which acquires an expectation value (not necessarily a *vacuum* expectation value) which is not invariant under the symmetry in question, we say that the system is in the ordered phase, and the symmetry is spontaneously broken. This is because other subsystems interact with the order parameter, which specifies a "frame of reference" to be measured against. In that case, the vacuum state does not obey the initial symmetry (which would keep it invariant, in the linearly realized Wigner mode in which it would be a singlet), and, instead changes under the (hidden) symmetry, now implemented in the (nonlinear) Nambu–Goldstone mode. Normally, in the absence of the Higgs mechanism, massless Goldstone bosons arise.

The symmetry group can be discrete, such as the space group of a crystal, or continuous (e.g., a Lie group), such as the rotational symmetry of space. However, if the system contains only a single spatial dimension, then only discrete symmetries may be broken in a vacuum state of the full quantum theory, although a classical solution may break a continuous symmetry.

A Pedagogical Example: The Mexican Hat Potential

In the simplest idealized relativistic model, the spontaneously broken symmetry is summarized through an illustrative scalar field theory. The relevant Lagrangian, which essentially dictates how a system behaves, can be split up into kinetic and potential terms,

$$\mathcal{L} = \partial^{\mu}\phi\partial_{\mu}\phi - V(\phi). \tag{1}$$

It is in this potential term $V(\Phi)$ that the symmetry breaking is triggered. An example of a potential, due to Jeffrey Goldstone is illustrated in the graph at the right.

$$V(\phi) = -10|\phi|^2 + |\phi|^4 \,. \tag{2}$$

This potential has an infinite number of possible minima (vacuum states) given by

$$\phi = \sqrt{5}e^{i\theta} \,. \tag{3}$$

for any real θ between 0 and 2π. The system also has an unstable vacuum state corresponding to $\Phi = 0$. This state has a U(1) symmetry. However, once the system falls into a specific stable vacuum state (amounting to a choice of θ), this symmetry will appear to be lost, or "spontaneously broken".

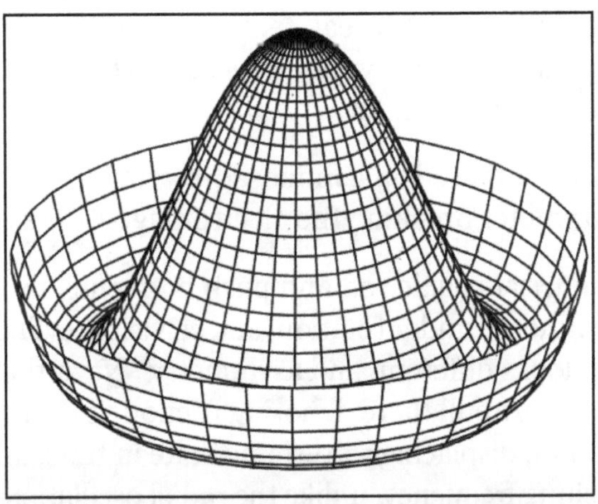

Graph of Goldstone's "Mexican hat" potential function V versus φ.

In fact, any other choice of θ would have exactly the same energy, implying the existence of a massless Nambu–Goldstone boson, the mode running around the circle at the minimum of this potential, and indicating there is some memory of the original symmetry in the Lagrangian.

Other Examples

- For ferromagnetic materials, the underlying laws are invariant under spatial rotations. Here, the order parameter is the magnetization, which measures the magnetic dipole density. Above the Curie temperature, the order parameter is zero, which is spatially invariant, and there is no symmetry breaking. Below the Curie temperature, however, the magnetization acquires a constant non-vanishing value, which points in a certain direction (in the idealized situation where we have full equilibrium; otherwise, translational symmetry gets broken as well). The residual rotational symmetries which leave the orientation of this vector invariant remain unbroken, unlike the other rotations which do not and are thus spontaneously broken.

- The laws describing a solid are invariant under the full Euclidean group, but the solid itself spontaneously breaks this group down to a space group. The displacement and the orientation are the order parameters.
- General relativity has a Lorentz symmetry, but in FRW cosmological models, the mean 4-velocity field defined by averaging over the velocities of the galaxies (the galaxies act like gas particles at cosmological scales) acts as an order parameter breaking this symmetry. Similar comments can be made about the cosmic microwave background.
- For the electroweak model, as explained earlier, a component of the Higgs field provides the order parameter breaking the electroweak gauge symmetry to the electromagnetic gauge symmetry. Like the ferromagnetic example, there is a phase transition at the electroweak temperature. The same comment about us not tending to notice broken symmetries suggests why it took so long for us to discover electroweak unification.
- In superconductors, there is a condensed-matter collective field ψ, which acts as the order parameter breaking the electromagnetic gauge symmetry.
- Take a thin cylindrical plastic rod and push both ends together. Before buckling, the system is symmetric under rotation, and so visibly cylindrically symmetric. But after buckling, it looks different, and asymmetric. Nevertheless, features of the cylindrical symmetry are still there: ignoring friction, it would take no force to freely spin the rod around, displacing the ground state in time, and amounting to an oscillation of vanishing frequency, unlike the radial oscillations in the direction of the buckle. This spinning mode is effectively the requisite Nambu–Goldstone boson.
- Consider a uniform layer of fluid over an infinite horizontal plane. This system has all the symmetries of the Euclidean plane. But now heat the bottom surface uniformly so that it becomes much hotter than the upper surface. When the temperature gradient becomes large enough, convection cells will form, breaking the Euclidean symmetry.
- Consider a bead on a circular hoop that is rotated about a vertical diameter. As the rotational velocity is increased gradually from rest, the bead will initially stay at its initial equilibrium point at the bottom of the hoop (intuitively stable,

lowest gravitational potential). At a certain critical rotational velocity, this point will become unstable and the bead will jump to one of two other newly created equilibria, equidistant from the center. Initially, the system is symmetric with respect to the diameter, yet after passing the critical velocity, the bead ends up in one of the two new equilibrium points, thus breaking the symmetry.

Nobel Prize

On October 7, 2008, the Royal Swedish Academy of Sciences awarded the 2008 Nobel Prize in Physics to three scientists for their work in subatomic physics symmetry breaking. Yoichiro Nambu, of the University of Chicago, won half of the prize for the discovery of the mechanism of spontaneous broken symmetry in the context of the strong interactions, specifically chiral symmetry breaking. Physicists Makoto Kobayashi and Toshihide Maskawa shared the other half of the prize for discovering the origin of the explicit breaking of CP symmetry in the weak interactions. This origin is ultimately reliant on the Higgs mechanism, but, so far understood as a "just so" feature of Higgs couplings, not a spontaneously broken symmetry phenomenon.

References

- R. Oerter (2006). The Theory of Almost Everything: The Standard Model, the Unsung Triumph of Modern Physics (Kindle ed.). Penguin Group. p. 2. ISBN 0-13-236678-9
- FermiFred (2010-02-15). "C.R. Hagen discusses naming of Higgs Boson in 2010 Sakurai Prize Talk". Youtube.com. Retrieved 01, March 2020
- Quantum Field Theory in a Nutshell (Second Edition), by A. Zee (Princeton University Press, 2010) ISBN 978-1-4008-3532-4
- "Confirmed: CERN discovers new particle likely to be the Higgs boson". YouTube. Russia Today. 4 July 2012. Retrieved 25, April 2020
- Close, Frank (2011). The Infinity Puzzle: Quantum Field Theory and the Hunt for an Orderly Universe. Oxford: Oxford University Press. ISBN 978-0-19-959350-7
- Steven Weinberg (20 April 2011). Dreams of a Final Theory: The Scientist's Search for the Ultimate Laws of Nature. Knopf Doubleday Publishing Group. ISBN 978-0-307-78786-6
- Mark Thomson (5 September 2013). Modern Particle Physics. Cambridge University Press. pp. 499–500. ISBN 978-1-107-29254-3
- Department of Physics and Astronomy. "Rochester's Hagen Sakurai Prize Announcement". Pas. rochester.edu. Retrieved 11, August 2020

Permissions

All chapters in this book are published with permission under the Creative Commons Attribution Share Alike License or equivalent. Every chapter published in this book has been scrutinized by our experts. Their significance has been extensively debated. The topics covered herein carry significant information for a comprehensive understanding. They may even be implemented as practical applications or may be referred to as a beginning point for further studies.

We would like to thank the editorial team for lending their expertise to make the book truly unique. They have played a crucial role in the development of this book. Without their invaluable contributions this book wouldn't have been possible. They have made vital efforts to compile up to date information on the varied aspects of this subject to make this book a valuable addition to the collection of many professionals and students.

This book was conceptualized with the vision of imparting up-to-date and integrated information in this field. To ensure the same, a matchless editorial board was set up. Every individual on the board went through rigorous rounds of assessment to prove their worth. After which they invested a large part of their time researching and compiling the most relevant data for our readers.

The editorial board has been involved in producing this book since its inception. They have spent rigorous hours researching and exploring the diverse topics which have resulted in the successful publishing of this book. They have passed on their knowledge of decades through this book. To expedite this challenging task, the publisher supported the team at every step. A small team of assistant editors was also appointed to further simplify the editing procedure and attain best results for the readers.

Apart from the editorial board, the designing team has also invested a significant amount of their time in understanding the subject and creating the most relevant covers. They scrutinized every image to scout for the most suitable representation of the subject and create an appropriate cover for the book.

The publishing team has been an ardent support to the editorial, designing and production team. Their endless efforts to recruit the best for this project, has resulted in the accomplishment of this book. They are a veteran in the field of academics and their pool of knowledge is as vast as their experience in printing. Their expertise and guidance has proved useful at every step. Their uncompromising quality standards have made this book an exceptional effort. Their encouragement from time to time has been an inspiration for everyone.

The publisher and the editorial board hope that this book will prove to be a valuable piece of knowledge for students, practitioners and scholars across the globe.

Index

www.ingramcontent.com/pod-product-compliance
Lightning Source LLC
LaVergne TN
LVHW080610210926
848485LV00003B/134
* 9 7 8 1 6 3 9 8 9 3 6 3 8 *